AF595008

Sustainable Rural Futures

Sustainable Rural Futures

Editor

John McDonagh

MDPI • Basel • Beijing • Wuhan • Barcelona • Belgrade • Manchester • Tokyo • Cluj • Tianjin

Editor
John McDonagh
National University of Ireland
Ireland

Editorial Office
MDPI
St. Alban-Anlage 66
4052 Basel, Switzerland

This is a reprint of articles from the Special Issue published online in the open access journal *Sustainability* (ISSN 2071-1050) (available at: https://www.mdpi.com/journal/sustainability/special_issues/rural_futures).

For citation purposes, cite each article independently as indicated on the article page online and as indicated below:

LastName, A.A.; LastName, B.B.; LastName, C.C. Article Title. *Journal Name* **Year**, *Volume Number*, Page Range.

ISBN 978-3-0365-4479-3 (Hbk)
ISBN 978-3-0365-4480-9 (PDF)

Contents

About the Editor

John McDonagh

John McDonagh is a rural geographer with interests in agriculture, small-scale farming, the multifunctional countryside and the challenges of rural sustainability. With an international reputation and an impressive track record of success in EU-funded Research Projects, John has published a number of books and a large number of articles in high impact journals, as well as delivering invited papers and keynote addresses across Europe and beyond. He received his PhD degree from Bristol University where he studied under Professor Paul Cloke, and he is currently Associate Professor and Head of Geography at NUI Galway (NUIG) in Ireland.

Preface to "Sustainable Rural Futures"

The future sustainability of rural areas is central to current discussions around food security, energy security and the challenge of combatting climate change. With its innate complexity, myriad contestations and often-fragmented policy strategies, the diversity of rural areas is pivotal in confronting the challenges posed by biodiversity loss, changing economic and social practices and the reconfiguration of global–local interactions. This book, written by experts with a global reach, presents recent research exploring issues around rural poverty, an ageing farming population, the challenges faced by young rural dwellers, issues of succession, generational renewal and farm viability to name just a few. In all, what is revealed here gives the reader fascinating insights into the contemporary debates and potential trajectories that our rural areas are currently contending with.

John McDonagh
Editor

Editorial

Rural Futures and the Future of the Rural

John McDonagh

Department of Geography, School of Geography, Archaeology & Irish Studies, National University of Ireland, Galway (NUIG), H91 TK33 Galway, Ireland; john.mcdonagh@nuigalway.ie

Citation: McDonagh, J. Rural Futures and the Future of the Rural. *Sustainability* **2022**, *14*, 6381. https://doi.org/10.3390/su14116381

Received: 18 May 2022
Accepted: 20 May 2022
Published: 24 May 2022

1. Introduction

To talk of rural and the future in the same sentence was described by Shucksmith [1] as being 'something of an oxymoron' (p. 163). Indeed, the rural has often been thought of and deliberated on in terms of its identifying with the past, fundamental to narratives around heritage, tradition, conservatism, perceptions of the rural idyll and a place in need of modernisation [2,3]. Currently, the rural is also at its most transformative period in our history. Centrally positioned in terms of its role in food production, energy security and most critically, confronting climate change, the rural finds itself portrayed as 'the fabric of our society ... the heartbeat of our economy ... a core part of our identity and our economic potential' [4]. While such sentiments are laudable, the rural also represents an amalgam of contradictions with some areas thriving and expanding, while others decline and become increasingly marginalised.

Accordingly, the ways in which rural areas engage with global challenges are multi-layered and numerous. Climate change and protecting biodiversity, for example, must be connected with and operationised at local levels as they have 'profound implications for the future use and regulation of rural space' (p. 28, [5]). Functioning within this space, the rural must also be cognisant of the diversity of rural stakeholders, objectives and key drivers such as urbanisation, globalisation, political and ideological pressures, and changing commodification practices that make up the countryside. These have a key impact on the future of rural places [6] and the ways in which 'conflicting and competing priorities around landscape protection and economic development' (p. 642, [7]) are dealt with.

The rural is also key to discussions on sustainability. The considerations here are often displayed through multitude rural policy programmes like those of the US and their sometimes 'fragmented and incoherent' [8] offerings, or the more reactive and often economically driven policies of Europe. One recent offering is the 'Long-term vision for Rural Areas' that espouses the need for stronger, connected, resilient and prosperous rural areas [4] throughout the European Union. What is less vocalized, however, is that the rural in Europe and elsewhere must be viewed against backdrops of ageing farming populations, limited land mobility, depopulation (particularly of rural youth), reduction in employment opportunities, disillusionment among young farmers in terms of future livelihoods, and a steady decline of the farm family. How to address these challenges into the future is considerable. The diversity of rural landscapes, traditions and cultures demands that any vision for rural sustainability must be able to 'incorporate people, practices, economies and environments that do not easily fit into the existing policy models and development visions' (p. 104, [9]). This will necessitate the need to strike some form of viable balance [10], whereby environment and economic goals are not mutually exclusive and where 'exploitative activities give way to an understanding of the complex ecosystems of which human economy and habitation are crucial parts' (p. 642, [7]).

Thus, when it comes to thinking about rural futures, complexities, contradictions and conflicts are many. Nevertheless, an ability to adapt, to strategize, to overcome unexpected obstacles and to build sustainable and resilient systems is a challenge and necessity that rural people and places frequently embrace. In this Special Issue collection, some of the ways in which this plays out on the ground are explored. Concerns around poverty, land use change and sustainability, rural regeneration, young rural dwellers, family farms, innovation and farm viability are delved into to provide insights and potential pathways which undoubtedly will be of use for future rural planning and future rural sustainability.

2. This Collection

The articles that make up this Special Issue negotiate a broad spectrum of questions and traverse an extensive geographical area. Authors from Canada, Spain, Scotland, Ireland, Mozambique, Hungary, Chile, Northern Ireland, Czech Republic and Slovakia bring interesting and thought-provoking commentaries in terms of both expertise and case studies.

The first article by Eldridge et al. (Contribution 1) delves into perspectives on the poverty trap and smallholder farmers in Tanzania. The essential question here is how smallholder farmers, essential in terms of food production and addressing poverty, remain 'trapped in a vicious cycle of endemic poverty' (p. 1). Focusing on the Meru district of Tanzania, the authors use the small-holder agro-input supply chain as their unit of analysis, allowing them to depict the complex and interconnected nature of smallholder farmers alongside insights into how and why the poverty trap both exists and persists. The key message is that primarily it must be acknowledged that smallholder farmers are crucial actors in global food production, and secondly, that in order for them to continue contributing in the way they do, it is necessary to develop 'sustainable livelihoods ... through reducing their susceptibility to the dynamics associated with poverty traps' (p. 29). It is only in addressing constraints around infrastructure, resources such as inputs, credit and information, and government policies and regulations, that such change might be realised.

The complex relationship between human well-being and land use change in Mozambique is the focus of the second paper in this collection. Here, Zorrilla-Miras et al. (Contribution 2) use a multi-scale participatory scenario planning process to explore pathways for agricultural, economic and social development, and their implications for changes in land use and land cover (LULC), ecosystems services and society well-being. The results that emerge from this scenario building exercise undoubtedly produce options for decision-makers in Mozambique, as well as providing a 'richer understanding and gains in context-specific knowledge on LULC and ecosystem services for human well-being (particularly) in areas with populations of vulnerable small-scale farms' (p. 20).

Young rural inhabitants is the theme of the third paper. Drawing from experiences in Chile, Rodriguez et al. (Contribution 3) explore place attachment and the threat to rural livelihoods and sustainability that exists from an exodus of young people to urban areas. While many studies focus on the forces that encourage young people to leave rural areas, such as education and employment, this paper identifies the opportunities that rural places offer. In particular, the authors advocate the need for local policy initiatives to realise the strengths of living in rural areas. They highlight those components that are important to young rural dwellers, such as a connection to nature, social constituents of living in the countryside, a sense of belonging and the importance of community. The paper contributes to our understanding of how these social relations, and relationships with the natural environment, play an important role in young people's appreciation and attachment to place, and how it influences where they want to live.

Kovach et al. (Contribution 4) in the fourth contribution explore what they call the 'unstoppable process', that is, the ageing farm workforce and the implications this has for a reduction in agricultural activity and consequent impacts on the landscape of the European countryside. Presented here is a fascinating insight into the complexity of challenges that young farmers, successors and new entrants into farming face. These include dealing with issues of education, access to land and family traditions, all of which ultimately influence future sustainability practices. In focusing on these, the paper opens up interesting discussions on the reasons that young people engage in farming alongside their ideas around sustainability and what that means in the context of their farming practice.

Attachment to land, a particular way of life, and the part played by working outside and with nature, were the typical responses elucidated by the young Hungarian farmers. Rather interestingly, what also emerged was their views on sustainability and what that meant in the context of their farming practice. The discussion here indicates that as well as having an environmental protection aspect, sustainability was very often couched in the economics of farming in that environmental schemes were engaged with, or in some way justified, only if it lead to either sparing money in terms of input materials or was useful in helping to access subsidies from Europe. The authors conclude that while nature conservation and environmental protection were important, there was a growing space for discussion around the need to 'strengthen the emergence of sustainability practices' (p. 12).

Holloway et al. (Contribution 5) in their contribution address one of the more sensitive aspects of future rural sustainability, namely that of how farms are passed on to the next generation. Exploring the emotional aspects of retirement and the succession decision-making process, the authors take us beyond the oft-used economic arguments to one that explores the emotional aspects that influence such decisions. What emerges from their innovative 'walk and talk' methodology is the complex relationship and emotional attachment farmers have to their farms, as well as the lack of appreciation and/or understanding policy makers have for such attachment. A greater understanding of emotional aspects, alongside that of a sense of place belonging, is identified as being crucial in the broader farm decision-making process, and particularly in the relationship between farmer and successor.

The sixth paper in this collection by Conway et al. (Contribution 6) continues with this theme by exploring ways in which the older farmer could be supported and reassured 'that their sense of purpose and legitimate social connectedness within the farming community will not be jeopardised upon handing over the farm business to the next generation' (p. 1). The need to overcome the typical succession processes of the past which is described as 'effectively obstruct(ing) the transfer of farmland from one generation to the next' can perhaps be tempered by addressing the needs of the older farmer more carefully. This, the authors argue, particularly relates to self-worth, farmer identity and quality of life. What emerges from this research is an interesting call for the development of a social organisation specifically for older farmers that would ensure a connection with their past farming lives remains in place. Such an organisation would undoubtedly impact the quality of life of those most impacted by successional change, and ultimately would help 'transform farming into an age-friendly sector of society' (p. 8).

Farrell et al. (Contribution 7) also use the family farm as the backdrop to their paper. Here, the focus is on options that may help ensure both viability and longer term sustainability. Recognising the role that the family farm plays in the broader social and cultural traditions of Europe, the attention here centres on exploring innovative practices that could play a role in encouraging younger farmers to become involved in farming, but which would also be important in terms of climate change, environmental protection, farm viability and ultimately long-term sustainability. The case study evidence is drawn from a group of Irish farmers engaged in the Maximising Organic Production System (MOPS) EIP-AGRI project. The subsequent empirical evidence gathered from interviews and focus groups suggest that not only would the uptake of diversified practices such as organics 'improve farm viability, but (it) would also encourage the next generation of young farmers to commit to the family farm and consider farming long-term'. (p. 1). The paper concludes with the assertion that opportunities presented by organic farming can be important in securing farm viability, and even more so can 'act as a catalyst in attracting new entrants to the agricultural sector' (p. 11).

In McDonagh's paper (Contribution 8), the discussion moves from the previous emotional arena whereby farmers who retire often find themselves on the outside looking in (Contribution 6), and the issue of farm viability and the pursuit of innovative practices towards sustainability (Contribution 7), to an arena drawing from both but which is often mired by contested and often conflictual engagements. Here, the discussion is built around farmers who, looking to enhance their farm's viability, often find themselves on the outside, not by virtue of retirement but by virtue of the top-down policies enacted on their lands under the banner of sustainable practices and environmental management. In this paper, the ways in which environmental objectives can be addressed alongside a farmers' ability to farm are the main considerations. The premise of the paper is that a top-down policy driven approach will not be successful, and what is required is an inclusive and farmer endorsed one. In particular, the argument is made that a combination of top-down and bottom-up with that of a locally-led vision can yield better environmental outcomes and more engaged farming practices. The 'key ingredients' that emerge from the research include the importance of multi-stakeholder involvement and a prominent role for farmers in the decision-making process. In all, the best path is a combination of action-based, results-based and locally-led programmes alongside the integrating of local and scientific knowledge in pursuit of the best environmental outcomes.

In the final paper of this collection, Vaishar et al. (Contribution 9) bring the discussion to Eastern Europe and the Moravian–Slovak borderland. In this paper, the possibilities for rural development in the context of the changing geopolitical positioning of the rural region of Eastern Moravia are explored. The paper considers the challenges posed for this region as it emerges from being marginal and agricultural to industrial, to its (re)positioning on the margins of the eastern border of Czechia. The 'movement' of borders and how issues such as migration and other indicators of mobility such as the construction of new dwellings provide insight to how this region is evolving. What emerges from the research is that East Moravia is on the threshold of change from manufacturing to a region of shared services with great opportunities for the rural, particularly in the context of cultural tourism development.

Collectively, the papers in this Special Issue make an interesting contribution to our understanding of the challenges facing rural areas across Europe and beyond. What is clear is the need to embrace the diversity that rural areas present, and allow this diversity to be reflected in policy discourse and practical application. The bringing together of farmer, policy maker and rural community is an important strategy and a powerful tool in shaping rural futures. Ultimately, with policy acting as a facilitator, the bringing together of diverse experiences, knowledge and resources will better equip the broader decision-making process and better enable sustainable rural futures.

3. Contribution List

1. Eldridge, E.; Rancourt, M.-E.; Langley, A.; Héroux, D. Expanding Perspectives on the Poverty Trap for Smallholder Farmers in Tanzania: The Role of Rural Input Supply Chains. *Sustainability* **2022**, *14*, 4971. https://doi.org/10.3390/su14094971
2. Zorrilla-Miras, P.; López-Moya, E.; Metzger, M.J.; Patenaude, G.; Sitoe, A.; Mahamane, M.; Lisboa, S.N.; Paterson, J.S.; López-Gunn, E. Understanding Complex Relationships between Human Well-Being and Land Use Change in Mozambique Using a Multi-Scale Participatory Scenario Planning Process. *Sustainability* **2021**, *13*, 13030. https://doi.org/10.3390/su132313030
3. Rodríguez-Díaz, P.; Almuna, R.; Marchant, C.; Heinz, S.; Lebuy, R.; Celis-Diez, J.L.; Díaz-Siefer, P. The Future of Rurality: Place Attachment among Young Inhabitants of Two Rural Communities of Mediterranean Central Chile. *Sustainability* **2022**, *14*, 546. https://doi.org/10.3390/su14010546
4. Kovách, I.; Megyesi, B.G.; Bai, A.; Balogh, P. Sustainability and Agricultural Regeneration in Hungarian Agriculture. *Sustainability* **2022**, *14*, 969. https://doi.org/10.3390/su14020969
5. Holloway, L.A.; Catney, G.; Stockdale, A.; Nelson, R. Sustainable Family Farming Futures: Exploring the Challenges of Family Farm Decision Making through an Emotional Lens of 'Belonging'. *Sustainability* **2021**, *13*, 12271. https://doi.org/10.3390/su132112271

6. Conway, S.F.; Farrell, M.; McDonagh, J.; Kinsella, A. 'Farmers Don't Retire': Re-Evaluating How We Engage with and Understand the 'Older' Farmer's Perspective. *Sustainability* **2022**, *14*, 2533. https://doi.org/10.3390/su14052533
7. Farrell, M.; Murtagh, A.; Weir, L.; Conway, S.F.; McDonagh, J.; Mahon, M. Irish Organics, Innovation and Farm Collaboration: A Pathway to Farm Viability and Generational Renewal. *Sustainability* **2022**, *14*, 93. https://doi.org/10.3390/su14010093
8. McDonagh, J. Designation, Incentivisation and Farmer Participation—Exploring Options for Sustainable Rural Landscapes. *Sustainability* **2022**, *14*, 5569. https://doi.org/10.3390/su14095569
9. Vaishar, A.; Šťastná, M.; Kramáreková, H. Moravian–Slovak Borderland: Possibilities for Rural Development. *Sustainability* **2022**, *14*, 3381. https://doi.org/10.3390/su14063381

Funding: This research received no external funding.

Conflicts of Interest: The author declares no conflict of interest.

References

1. Shucksmith, M. Re-imagining the rural: From rural idyll to Good Countryside. *J. Rural. Stud.* **2018**, *59*, 163–172. [CrossRef]
2. Ward, N.; Ray, C. *Futures Analysis, Public Policy and Rural Studies*; CRE WP74; Centre for Rural Economy, Newcastle University: Newcastle, UK, 2004.
3. Shucksmith, M.; Brown, D.; Shortall, S.; Vergunst, M.; Warner, M. *Rural Policies and Rural Transformations in the US and the UK*; Routledge: New York, NY, USA, 2012.
4. European Commission (EC). Available online: https://ec.europa.eu/info/strategy/priorities-2019-2024/new-push-european-democracy/long-term-vision-rural-areas_en (accessed on 6 May 2022).
5. Woods, M. Grounding Global Challenges and the Relational Politics of the Rural. In *The Sustainability of Rural Systems—Global and Local Challenges and Opportunities*; Cawley, M., Bicalho, A., Laurens, L., Eds.; CSRS of the IGU and Whitaker Institute NUI Galway: Galway, Ireland, 2013; pp. 27–35.
6. Woods, M. The future of rural places. In *The Routledge Companion to Rural Planning*; Scott, M., Gallent, N., Gkartzios, M., Eds.; Routledge: Abingdon, UK, 2019; pp. 622–623.
7. Scott, M.; Gallent, N.; Gkartzios, M. Planning and rural futures. In *The Routledge Companion to Rural Planning*; Scott, M., Gallent, N., Gkartzios, M., Eds.; Routledge: Abingdon, UK, 2019; pp. 633–644.
8. Pipa, A.; Geismar, N. Reimagining rural policy: Organizing federal assistance to maximize rural prosperity. Available online: https://www.brookings.edu/research/reimagining-rural-policy-organizing-federal-assistance-to-maximize-rural-prosperity/ (accessed on 12 May 2022).
9. Mincyte, D. Subsistence and Sustainability in Post-industrial Europe: The politics of small-scale farming in Europeanising Lithuania. *Sociol. Rural.* **2011**, *51*, 101–118. [CrossRef]
10. Varley, T.; McDonagh, J.; Shortall, S. The Politics of Rural Sustainability. In *A Living Countryside? The Politics of Sustainable Development in Rural Ireland*; McDonagh, J., Varley, T., Shortall, S., Eds.; Ashgate: Farnham, UK, 2009; pp. 1–24.

MDPI

Article

Expanding Perspectives on the Poverty Trap for Smallholder Farmers in Tanzania: The Role of Rural Input Supply Chains

Elizabeth Eldridge [1,*], Marie-Eve Rancourt [1], Ann Langley [2] and Dani Héroux [1]

[1] Department of Logistics and Operations Management, HEC Montréal, Montréal, QC H3T 2A7, Canada; marie-eve.rancourt@hec.ca (M.-E.R.); dani.heroux@hec.ca (D.H.)
[2] Department of Management, HEC Montréal, Montréal, QC H3T 2A7, Canada; ann.langley@hec.ca
* Correspondence: elizabeth-anne.eldridge@hec.ca

Abstract: Smallholder farmers across rural landscapes remain trapped in a vicious cycle of endemic poverty where interconnected challenges limit their ability to improve their livelihoods. Our study of smallholder farmers' relationships with suppliers and several stakeholders across the Tanzanian rural agro-input supply chain offers an extended perspective on the persistence of endemic poverty and broadens the discussion on the future of sustainable food production and smallholder livelihoods. Through interviews and focus groups, we use a grounded theory methodology to develop a systemic approach to understanding the complexities of this landscape as related to smallholder agro-input sourcing activities. Our causal loop diagram framework provides a unique perspective on the poverty trap experienced by smallholder farmers in this context. Our findings may be useful in targeting practical and sustainable directions towards overcoming the poverty trap, ultimately enabling smallholders to increase wealth and improve their livelihoods through sustainable practices.

Keywords: smallholder farmer; input sourcing; Tanzania; poverty reduction; grounded theory; rural agriculture

Citation: Eldridge, E.; Rancourt, M.-E.; Langley, A.; Héroux, D. Expanding Perspectives on the Poverty Trap for Smallholder Farmers in Tanzania: The Role of Rural Input Supply Chains. *Sustainability* **2022**, *14*, 4971. https://doi.org/10.3390/su14094971

Academic Editor: John McDonagh

Received: 3 March 2022
Accepted: 13 April 2022
Published: 21 April 2022

Publisher's Note: MDPI stays neutral with regard to jurisdictional claims in published maps and institutional affiliations.

1. Introduction

Smallholder farmers in developing countries play an essential role in food supply chains, where over 70% of global food requirements are generated by small-scale producers [1]. However, they remain overwhelmed by endemic poverty, often living on less than $2 per day [2]. It is puzzling that these major contributors receive so little value for their efforts. Their entrenchment in poverty poses a risk to their own subsistence while directly (or indirectly) supporting the subsistence of others. This trend is common across Sub-Saharan Africa, including Tanzania as our area of study.

Poverty, poverty traps and the factors influencing them are a well-studied and arguably controversial topic [3–6]. Although multiple causes, effects, and solutions have been offered across a wide variety of disciplines, there is still room for further research into their origins. Notably, our study offers a unique perspective by focusing on input supply chain components and processes as the unit of analysis. This enables us to offer novel insights into the nature of the endemic poverty trap facing smallholders and how the risks to their subsistence are perpetuated and reinforced in problematic supply chain interactions.

Our study aimed to develop a deep understanding of the current environment experienced by smallholders rather than testing an a priori hypothesis. Over an intensive one-month period we conducted a qualitative grounded theory study of the agricultural crop input supply chain of Meru District, Tanzania, a setting that is likely typical of conditions in a broader range of regions and countries where smallholder populations encounter similar challenges, thus offering the possibility for broader application of our findings. Our study aimed to address two research questions: (1) How does the organization of smallholder rural agro-input supply chains contribute to the poverty trap for smallholder farmers? (2) How might the challenges experienced by smallholders be overcome to promote sustainable rural livelihoods?

A "poverty trap" is defined as a self-reinforcing situation where those affected by poverty are unable to escape; in essence, poverty begets poverty which can be brought on by a variety of macro and micro stressors [3–6]. Poverty traps are problematic in that they lock smallholders into their current position where opportunities to improve on-farm productivity necessary for growth, increased revenue, and investment into more sustainable technologies, inputs, and practices may not be readily available or accessible. In fact, the United Nations had identified ending poverty as the first of 17 Sustainable Development Goals and as crucial to sustainable development, with their second goal (Zero Hunger) being, in part, oriented towards promoting sustainable agriculture [7].

Notwithstanding research that explores the diverse aspects of poverty traps, market dynamics and associated smallholder participation, and the existing body of literature regarding inputs, [3,8–16], we see a need for more in-depth investigation into supply chain dynamics with a particular focus on the challenges that are still restricting smallholders from accessing improved inputs, how this contributes to the poverty trap, and how these challenges may be overcome. This aligns with the premise presented by Chambers and Conway [17] where they note the importance of an integrated and holistic approach to the discussion of sustainable rural livelihoods with capability, equity, and sustainability as core concepts given the complexities associated with rural development. Using supply chains as the basis to conduct analysis on the poorest socio-economic demographics—particularly within rural communities in developing countries—can offer a valuable link between two very different research streams and promote sustainable and practical solution spaces for the various stakeholders involved [18,19].

When characterizing the smallholder agricultural environment within a supply chain nexus as being a feedback system comprised of an input component (i.e., materiel, equipment, and resources required to produce agricultural outputs) and an output component (i.e., market), we note that many scholars have placed emphasis above all on the challenges smallholders experience in relation to their market-related activities as one cause of endemic poverty [14–16], explaining how poverty traps may be generated and propelled. In this study, we instead focus on the activities associated with the smallholder agro-input supply chain, which we argue plays as significant a role within the system as the market in terms of its contributory effects on smallholder poverty, even before market considerations come into play. However, smallholder farmer input supply chains, and their effects on the poverty trap, have not received the attention they deserve in the literature.

Our system-based approach offers two main contributions. First, we examine the rural agro-input supply chain itself, providing insight into the experience of smallholders and allowing us to present an original conceptual framework in the form of a causal loop diagram reflecting the complexity of their challenges. Second, we suggest areas where targeted and collective action could be taken (e.g., by governments, nongovernment organizations (NGOs), and industry) to improve the lives of smallholders over the long term.

Following the literature review and description of our methods, we present a detailed review of supply chain participants and then progressively develop of our causal loop model of the input supply chain poverty trap drawing on our data. Finally, we describe ongoing efforts to address the challenges and opportunities for sustainable action suggested by our analysis.

2. Literature Review

2.1. Agricultural Crop Inputs in Sub-Saharan Africa

Several Sub-Saharan countries experience low agricultural productivity, which, according to the literature, is linked in part to the "inadequate use of modern inputs" [8,9,13]. As such, it is recommended that modern inputs, such as fertilizers, improved seeds, and various chemicals, be used more frequently to counter the challenge of low productivity and provide higher profits to farmers [20]. For example, the use of and understanding of inorganic fertilizer application has been shown to offer a solution to the productivity challenge [21,22] and can be used in combination with improved seeds to aid in increasing crop yields, eliminating poverty, and improving food security [23] while generating resilient incomes [24]. Additionally, the use of agro-chemicals (i.e., herbicides) can contribute to increased crop yields and can compensate for a deficit in human labour and challenges associated with soil erosion in the agricultural sector while also offering food security benefits [25,26]. Through increased use of improved inputs, there can also be a reduction in the production and operating costs and an improvement to planting and harvesting timelines. However, to achieve this, inputs must be sourced in an effective manner and used correctly [27] with adequate follow-up and monitoring [23]. This can then enable agricultural growth, which would promote regional economic development and lead to poverty reduction [28].

With so many potential benefits from improved inputs, one might ask why they are not being used at every opportunity by smallholders. Snyder et al. [29] suggests that, although there is a focus on technical aspects of yield gap analyses, the broader social, political and environmental context which may encourage or discourage farmers from making decisions and taking action is often ignored [30]. Although other studies have noted a wide range of possible factors that can impede the adoption of improved inputs, they tend to focus on a single factor or a small number of factors including issues such as capital, cost, forecasting, and supplier distribution [8,22,23,31,32], rather than the full spectrum of interconnected challenges. This does not lend itself to presenting a complete picture of the complex environment in which decisions are made. Furthermore, while cooperatives, larger farmer organizations, and NGOs have attempted to overcome smallholder problems of access to input markets, this issue remains prominent across the smallholder environment. By using a wider lens and perspective, as opposed to focusing on a particular crop, input, or explanatory factor, our research provides insights into the complex set of reasons why improved input use remains a challenge.

2.2. Two Sides of the Equation and the Missing Link: Contributions to the Poverty Trap

Research has noted the importance of market-based, downstream activities in poverty reduction, and efforts have been focused towards this area in the literature, where the smallholder is the supplier of crops. Barrett [16] suggests that market participation is the key to allowing smallholders to escape poverty, by generating sustainable income and encouraging more general economic growth [14,15]. Thus, a reduction in the costs of accessing markets, better organization of smallholders and improved access to production resources would benefit smallholders [16]. Much policy research has also been conducted on how to encourage smallholder market participation [33–36]. The literature has also focused on various aspects of smallholder market-based decision making, such as whether to send crops immediately to market or store crops post-harvest to gain potential benefit from later sales [37].

Given that the current research on market activities affecting farmers in developing countries is more extensive, there are also suggestions of practical and policy-driven mechanisms through which market relationships can be addressed. For example, market-enabling activities such as fair-trade [38] and certified organic production are gaining traction in some local markets across Africa [39]. However, these are primarily oriented towards exports and larger farming operations and are therefore not necessarily accessible to smallholders producing non-export crops, particularly in Sub-Saharan Africa [40]. Contract farming is another example. This activity is based on an agreement between the producer and the buyer with stipulations on product quantity, quality, delivery, and price, whereby the buyer often provides inputs and other resources to the producer. Considerable research has been conducted in this area [41–47], and contract farming may offer one mechanism to link the input and output components of the supply chain [41]; however the benefits and circumstances under which welfare may be improved remain unclear [48–50]. Some literature points towards contract farming providing an opportunity to access quality inputs as well as a guaranteed buyer, and opportunities to reduce income variability, improve household welfare, and promote risk transference [51]. However, this does not offer a holistic solution to the various other challenges smallholders face when sourcing inputs and does not necessarily consider the complexity of smallholder input sourcing across rural environments. Further, while contract farming may offer opportunities for technology adoption and increased productivity, as well as the potential for smallholders to grow out of dependency, the literature suggests that they may also grow further into it from a market perspective and that contract farming does not always contribute to profit generation, increased income, and poverty reduction [47,52]. In some cases, neither fair trade nor contract farming offer a positive contribution to sustainable development or to the improvement of smallholder livelihoods. Indeed, contract farming may contribute to continued poverty or even poverty trap-generation [53,54], where there is a risk of self-exploitation in the case of contract farming [55].

Poverty traps may occur for a variety of reasons, such as unique constraints based on scarcity of resources which force certain decisions that reinforce poverty, weak policy and institutional factors, economic activity, various external factors, low risk tolerance, etc. [3,10]. Additionally, many types of poverty traps may exist and range from the macro (country-level) to the micro (individual or household level) [11,12], the latter of which will be our focus in the paper. While there is a range of discussion on poverty traps and their contributing factors, they are often viewed through a purely economic lens. Existing studies may not necessarily fully incorporate some of the important contexts that define rural and smallholder demographics which can have negative impacts on livelihood sustainability [56,57]. Furthermore, given the abundance of ways by which poverty traps may occur, and the variables that contribute to them, there is a need to continue to investigate their underlying causes and mechanisms [58]. As such, a study such as ours that analyzes the issues surrounding input sourcing within the agro-input supply chain and that incorporates the core concepts detailed by Chambers and Conway [17] can supplement existing market-based analyses as part of a more comprehensive and broader approach, in order to better understand why poverty persists in these communities.

2.3. *Captivity, Risk and Power*

Relationships are an important aspect of any supply chain and are particularly important within the context of smallholders, given their propensity to use informal, trust-based contracts [59]. While operating in an informal environment, gaps between individual expectations of accountability and transparency [60] result in varying levels of control within each relationship which depends on the relative power of each actor. Ultimately this can result in a series of complex relationships with varying levels of risk for every participant in the process [61]. Bensaou [61] describes the situation of buyer/supplier captivity, where one actor finds him/herself a captive buyer to one or a few established suppliers who wield greater bargaining power within a concentrated market defined by stable demand, minimal innovation, and limited growth. Bensaou [61] also notes that captive suppliers can be found in unstable markets with high competition and few buyers, leading suppliers to be heavily dependent on their buyers, and with reduced bargaining power.

The concept of captivity implies dependency, reflecting the tenets of Resource Dependence Theory (RDT) where high dependence on other actors (e.g., buyers and/or suppliers) for needed resources, and the apparent absence of alternatives, can create a trap from which it can be difficult to escape [62]. RDT offers an important theoretical lens for understanding this challenge. In particular, by focusing on supplier relationships, we can better understand the risks associated with dependency [63]. Power and dependency are inextricably linked within RDT where buyers' and suppliers' mutual dependency on each other confers power [64,65]. Where one partner controls access to resources for the other, they are placed in a dominant position in the relationship [66,67].

Thus, resource dependency can pose significant risks for actors in the supply chain who have few alternatives. Indeed, in the supply chain literature, risk is often presented in terms of supplier or customer relationships, or as internal and external challenges, and it is divided into various categories based on the drivers which define the events and conditions leading to potential losses and negative impacts on operations. These categories of risk posed to businesses include disruptions, delays, technological systems, demand forecasting, intellectual property, procurement, receivables, inventory, and capacity [68]. Within agricultural supply chains, there may be other risk categories which include risks related to weather, environment, disease, sanitation, natural disasters, markets, logistics and infrastructure, management and operations, public policy and institutions, and politics [69]. It is understood that smallholder farmers may be particularly vulnerable to these risks and that this can be a contributing factor to the poverty trap [3,10].

We propose that another risk category should be added to the discussion of smallholder farmers and poverty traps—the risk of subsistence, meaning their ability to survive. This is a culmination of the other risk factors, where smallholders face an existential threat which is fed by other drivers or sources of risk. To illustrate this, Valkila [54] suggests that within Fair Trade arrangements, any extra income is often not enough for smallholders to feed their families, let alone to provide the opportunity to increase production, expand activities, or buy land. In some cases, by participating in more integrated markets such as Fair Trade, the risk may be shared across multiple stakeholders; however, with smallholders who do not or cannot participate in these integrated markets, their risk becomes much higher with much lower returns [70]. Furthermore, Livingston et al. [70] note that risk management stemming from the trade-offs that take place in terms of risk–reward is the key challenge facing smallholders in their ability to increase investments that would support increased productivity.

When we look at smallholder positioning within a rural supply chain, we see that their relationships with other actors could lead them into being both a captive buyer and a captive supplier; however, the dynamics by which this happens in specific cases is not clear a priori. Our study identifies and examines these issues based on the specific situation of smallholder farmers in rural Tanzania.

3. Materials and Methods

3.1. Research Context

This study focuses on Tanzania, a Sub-Saharan African country whose economy has a high dependence on agriculture, constituting 65% of the workforce and slightly less than one quarter of GDP [71]. This essential sector is predominantly comprised of smallholder farmers who are responsible for approximately 75% of total agricultural output [72]. Despite their high value to the economy, 39% of smallholders find themselves below the national poverty line [72]. Limited access to modern inputs results in low productivity, variable yields, and low profits [73]. This contributes to ongoing poverty, making Tanzania a highly suitable context for our study.

Our fieldwork was conducted in the Meru District of Tanzania (Figure 1) over a one-month period, in partnership with Farm Radio International (FRI), a Canadian NGO that uses radio to strengthen farming communities by partnering with local radio stations to broadcast information focusing on agriculture and rural development throughout Sub-Saharan Africa.

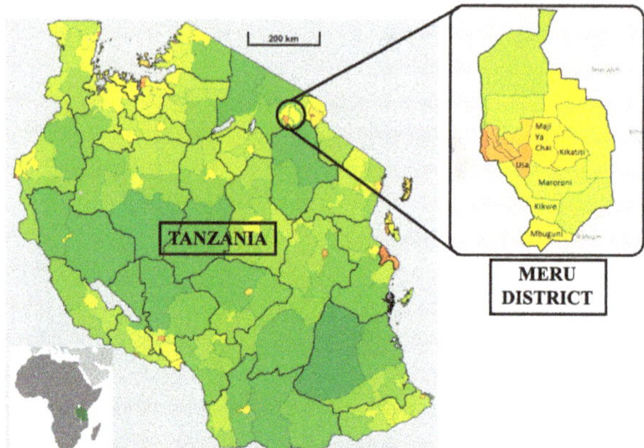

Figure 1. Geographical Layout of Fieldwork [74,75].

3.2. Research Design and Sampling

We follow an exploratory naturalistic inquiry research design, focusing on the experiences of people within their social and cultural contexts [76]. A grounded theory methodology [77] was employed based on a systematic process of constant analysis and comparison of data derived from the participants' experiences, through which we aimed to develop theory rather than test an a priori hypothesis. Initial, purposeful sampling [77] began by selecting villages where we could connect directly with smallholder farmers, and multiple suppliers were contacted for individual interviews. The overall sampled group at this stage included male and female smallholders of all age groups, village-level (local) suppliers, large-scale suppliers, and Agricultural Extension Officers (AEOs) at the district and village level. Over the course of the initial sampling, some gaps were identified, which necessitated further theoretical sampling [77] to include smallholder farmers, a large-scale supplier and a Tanzanian national farmer organization. Table 1 provides composition details across both samplings which took place over an intensive one-month period (November–December 2019).

Table 1. Sample Composition.

Type	Location	Interview Type and Participant Data	Objective
Initial sampling	Kikatiti	Focus group—Smallholder farmers (Male and female, 20 participants)	Understand how smallholders source and obtain their inputs, and the associated challenges. Determine the critical inputs. Explore possible solutions to challenges.
	Kikatiti	Focus group—Smallholder farmers. (Male and female, 8 participants)	
	Kikatiti	Focus group—Smallholder farmers (Male and female, 9 participants)	
	Kwaugoro	Focus group—Smallholder farmers (Male and female, 33 participants)	
	Mbuguni	Focus group—Smallholder farmers (Male and female, 18 participants)	
	Usa River	Village AEO (Female) District AEO (Female)	Determine the existing regulations and external conditions.

Table 1. *Cont.*

Type	Location	Interview Type and Participant Data	Objective
	Kikatiti	Village-level ago dealer/supplier (Male)	Understand activities further upstream. Determine challenges and potential solutions. Explore issues that smallholders identified and validate.
	Maji Ya Chai	Village-level ago dealer/supplier (Female)	
	Arusha	Importer/producer/distributor (Male)	
	Arusha	Importer/producer/distributor (Female)	
Theoretical Sampling	Kikwe	Focus group—Smallholder farmers (Male and female, 16 participants)	Revisit the initial smallholder statements using different techniques and understand how smallholders make decisions.
	Karangai	Focus group—Smallholder farmers (Male and female, 12 participants)	
	Arusha	Email interview—Meru Agro	Revisit the Meru Agro Lead Farmer Initiative
	N/A	Email interview—National Farmer Organization	Discuss the roles and contributions

3.3. Data Collection

Data were collected through semi-structured focus groups and interviews and was facilitated by FRI's Tanzanian Office. In total, data were collected from seven focus groups spanning five different villages in Meru District, for a total of 113 participants. Within one focus group, three participants representing local village leadership were also present. Two local suppliers and two large-scale suppliers were interviewed, as well as two AEOs. Email correspondence allowed us to re-interview one of the original respondents and to engage with a representative of a national farmer organization, resulting in data being collected from a total of 123 participants. Each focus group and interview was audio-recorded, translated (as required), and transcribed, leading to 124 pages of transcribed fieldnotes, and 81 pages of translated and transcribed audio files.

3.4. Data Analysis

The data were coded using procedures suggested by Charmaz [77] and Gioia et al. [78]. We began with in vivo (first order) coding which remained very close to the data, followed by higher level axial and theoretical coding (developed from the initial codes) that is more conceptual and captures larger segments of data. To show how initial in vivo codes were grouped together to extract more abstract themes, we provide a data structure diagram that illustrates our first order concepts, second order themes, and aggregate dimensions [78] building cumulatively on each other. These labels are the terms we use throughout the description of our findings.

Our fieldnotes provided the platform for our coding process, data analysis, and subsequent development of our data structure. Table 2 presents some coding samples and Table 3 presents our data structure. Over 600 separate concepts were identified through a line-by-line analysis of responses, which were further analyzed and grouped into ten second-order themes. Our aggregate dimensions were developed through an analysis of the frequency of appearance of second order themes across all the participant groups. The three most frequently occurring themes provided a starting point to develop our dimensions. The remaining seven themes were deemed sufficiently important to be retained; five of these were linked to one of the top three themes, based on similarity and relevance. The remaining two themes were relevant and significant to each other and thus contributed to the development of a fourth distinct dimension.

Table 2. Coding Samples.

Group	Associated Quote	Theme	Dimension
Smallholder Farmer	"For example, [you] worked hard all season, put a lot of expense [into] farming, and at the end of the day you don't get a good price for your crops. But also, you have a lot of needs. [...] I need to send my kids to school and in the middle of the season I have to pay [back some loans], [...]. Once you have your crops, you will need to sell even if it's [at] cheap price, you have no choice. You cannot wait until the price gets higher. [...]".	Captivity from Buyers and Suppliers	Unequal Power Dynamics
Smallholder Farmer	"The only reason why our crops go bad before it goes to the market [is] because we don't have modern machines to keep them fresh. [...]".		
Smallholder Farmer	"So, what we do, we just look to the supplier, so when we see a supplier selling more than, most people go to that shop. We just go there. We [assume] maybe his seeds are the best seeds or he has good quality and stuff like that, so that's why we go to that shop when we see many people buying from that shop, so we can go to that seller to buy our seeds".	Subsistence Risk	
Smallholder Farmer	"Sometimes you do have 20 sacks of maize and you want to keep them until the price gets high, but you can't do that because you don't have money to buy chemicals or pesticides to keep the maize in good condition. At the end of the day, you have to take your maize outside. You use the sun, and sometimes you don't have money also to buy something to cover in case of rain. It's a really big challenge, at the end of the day, you have to sell your maize or whatever you have for a cheap price".		
Large-scale supplier	"I have the statistics that show that improved seed in Tanzania is 18%. 18% of smallholder famers use improved seed over the last year. Everybody else will use farmer-saved seeds".	Availability Of Resources and Quality Inputs	Accessing Resources and Quality Inputs
Smallholder farmer	"The women in our village, they struggle to get fresh drinking water. They travel a long way to get water and at the end of the day they don't get time to go to the farm. The whole village here, we don't have water so that's a big, big challenge for us".		
Large-scale supplier	Because I go direct to a farmer [and] I make sure that farmer gets what is from me directly. Nothing has happened in between. So you can be able to trust [our] inputs.		
Smallholder farmer	"Nowadays [...] there's a lot of fake seeds and fertilizers, which is driving us crazy as farmers".		

Table 2. *Cont.*

Group	Associated Quote	Theme	Dimension
Smallholder farmer	"Before [an NGO] came over here, we used local transport like motorcycles, bicycles or walking. Myself, I use a donkey to go get the seeds from the center to my farm, but nowadays, since [the NGO] came over, they bring the seeds and all the stuff close to our village, so we don't have to go far away to get the agricultural [inputs] anymore".	Physically Accessing Inputs	Access to Information and Support
Smallholder farmer	"Well, we do have this challenge sometimes. The problem is we have lack of transport and lack of infrastructure. Some of the road in the village here are not good. But we try to deal with the challenges. Sometimes we organize the whole village to repair the roads so we can get our crops off the farm. Sometimes, it's really difficult to get the crops from the farm, sometimes you don't have any transportation to get them out. So, at the end of the day, they just go bad and you lose your crops, some of them. So, it's really, really challenging us".		
Smallholder farmer	"But some of the farmers, they don't have that education so they don't know which is fake and which is original, because they don't really even look at the package label and read them. But some of us, we have opportunity to get education from organizations and nowadays we know how to look for the quality and how to read the label, and to get to know which is fake and which is original".	Access to Information and External Support	
AEO	"We advise them to buy before, preparations. It's very important to buy them before the season". "One month before" "Most of them wait for the rain to come". "They are not sure of [when] the rain [will come], [this is why] we encourage them [to buy inputs ahead of time".		
AEO	"A big challenge [is] capital. Because [with] pesticides, fertilizers, you can see the price is increasing. So, a farmer cannot afford to buy all the inputs necessary, necessary inputs. But you can find, a few farmers who can afford it, but the others cannot. But we as extension officers, we advise them to, connect them to banks to get loans and the other institutions, you know, that get capital. Also, in the village, we advise farmers to start [Village Community Banking Associations]. "	Relying on Others	
Smallholder farmer	"To be honest, when it comes to check the quality of the product, it's really difficult for most of us, because most of us are not educated. So, it's really difficult, it's a big challenge for us because we don't know. Some of us, [we] don't know how to read and that's the problem".		

Table 2. *Cont.*

Group	Associated Quote	Theme	Dimension
Smallholder farmer	"It depends with the season. Sometimes there is long season and short season. We don't really get the seeds or fertilizer before the rain starts. So, when the rain starts, we get to know that this is the short season or long season. Normally on our side the rains start in February up to April, but sometimes the rain can start in March. Once the rains start in March, you really know the season will be short. So, I have to go to the shop and buy seeds for the short season. So, that's why we wait for the rains and the season to start so you really get to know if the season will be long or short".	Forecasting, Planning, and Preparation	Trade-offs and Decision Making
Smallholder farmer	"Overall, seeds, we don't get seeds at the right time. Sometimes the season gets started and there's no seeds because the supplier of the seeds, they don't really make sure that the seeds are there at the right time. We have this problem; we don't get seeds at the right time".		
Smallholder farmer	"Definitely, if we knew that there [are] original [/not fake] seeds and overall agricultural equipment and stuff like that, we would definitely organize ourselves as a village, and go there, get the seeds and all the equipment we need. Because we know that at the end of the day, we're going to benefit because that stuff is original".	Wanting to do Better	
AEO	"When you talk of different regions, [in the] Southern Highlands, there are a lot of farmers who grow maize...the Tanzanian government, they buy those crops, maybe for example maize, they buy them if there is in excess. So, [smallholders] plant it, they grow most crops, especially maize. If [there] is surplus, the Tanzanian government buys [the] maize and puts it in national food reserve to ensure food security in our country. If it happen[s] that the nearby country, maybe they have deficit of food, [the] Tanzania government [sells] the food to other countries. They have been divided into zones. [In the] northern zone we have national food reserve, Southern highland, [there are] two or three [National Food Reserves]. In our case, we don't know [how much is paid to smallholders]".		
Smallholder farmer	"The thing is that, when the season starts and sometimes, the season has started but you don't find seeds, most of the time, especially seeds like beans. You can go to the shop but you don't find seeds at the right time, that's the problem".	Ad hoc Decision Making and Pressure to Make Trade-offs	
Smallholder farmer	"The issue is that they need money during that time, so they have to sell. They have to sell".		

Table 3. Data Structure.

First Order Concepts	Theme	Dimension
▪ Relying on yields for survival ▪ Being held captive by the market/Saturated market reduces selling price ▪ Sourcing occurs through limited channels ▪ Purchasing options are limited by availability of input and limited supply ▪ Farming is their only experience/Not having another choice other than to farm ▪ Encouraging a cycle of poverty	Captivity from Buyers and Suppliers	Unequal Power Dynamics
▪ High opportunity cost for the farmer, no impact to suppliers (local or large-scale) ▪ Spending limited funds without certainty of return on investment ▪ Returning inputs is a time-consuming processes/Possibility for reimbursement is supplier-dependent/Risking delaying the planting season ▪ [Suppliers] risking consequences if regulations are not adhered to ▪ Risking high inventory holding costs [suppliers purchasing at wholesale quantity] ▪ Being exposed to theft (of packaging)	Subsistence Risk	
▪ Lacking capital/Financing options and payment mechanisms are limited ▪ Limiting quantity and type (quality) of products of that be purchased ▪ Not making enough to buy inputs for the next season/Production costs are higher than sales ▪ Lacking adequate storage for inputs ▪ Not enough time to do everything—investigating different avenues for input sourcing is not a top priority ▪ Encouraging alternative sourcing options and cooperative solutions ▪ Doing the most possible to ensure quality when purchasing ▪ Not knowing true quality before planting/Room is being left in the supply chain to tamper with inputs ▪ Trying to increase yields by using improved inputs ▪ Relying on unreliable/reduced quality self-harvested seeds ▪ Being held to a national standard for quality ▪ Reducing time in inventory/the time inputs are on shelves	Availability of Resources and Quality Inputs	Accessing Resources and Quality Inputs
▪ Carrying capacity is limited ▪ Travelling to suppliers is time-consuming and expensive ▪ Lacking transportation options ▪ Poor road conditions/infrastructure impedes access to inputs ▪ Preferring local suppliers due to accessibility ▪ Smallholders accessing large-scale suppliers and quality products is possible	Physically Accessing Inputs	

Table 3. *Cont.*

First Order Concepts	Theme	Dimension
▪ Changing government involvement/Government understaffing directs responsibility to uninvested players ▪ Limited understanding can result in barriers for farmers/Needing to know how to gather and use information effectively ▪ Lacking efficient flow of information downstream ▪ Accessing information via NGOs impacts relationship development with extension officers ▪ Inconsistent knowledge amongst farmers/Not knowing how to approach finding solutions ▪ Monitoring suppliers/Auditing is not having the desired results ▪ No follow-up/Minimal follow-up ▪ Providing limited scope solutions/Not tailoring solutions to end-user needs	Access to Information and External Support	Access to Information and Support
▪ [Smallholders] relying on suppliers to validate quality and provide quality products ▪ [Smallholders] relying on supplier for information and education ▪ [Smallholders] relying on nongovernment agencies for support ▪ [Local Suppliers] receiving training but relying on certifications from [Large-scale] supplier ▪ [Extension Officers] relying on farmers to share learned information leaves gaps in communication ▪ [Suppliers] relying on customers [Smallholders] to identify issues with the product/Relying on customers to ask questions	Relying on Others	
▪ Making trade-offs between cost and quantity purchased/Prioritizing price over quality ▪ Taking away resources from the family that were not intended for sale ▪ Coordinating efforts alone/without guidance ▪ Feeling afraid of negative repercussions from suppliers/Mistrusting suppliers ▪ Feeling desperate for money/Focusing on prioritizing basic activities such as water collection ▪ Feeling uncertain (quality, how to use inputs, prices, etc.)/Feeling pressured and overwhelmed	Ad Hoc Decision Making and Pressure to Make Trade-offs	Trade-offs and Decision Making
▪ Desiring to increase income ▪ Need to find ways to change what they are doing, not where they are doing it ▪ Wanting to be more in control/, to change, to save time and money, and to have recourse options ▪ Willing to change processes/Willing to try new approaches to input sourcing ▪ Willing to work more/longer for increased payoff ▪ Working together and merging resources is the key to success	Wanting to do Better	

Table 3. *Cont.*

First Order Concepts	Theme	Dimension
▪ Reacting vice being proactive ▪ Reacting to external and uncontrollable variables (i.e., weather) ▪ Sourcing begins under a time-constraint/Buying whatever is available ▪ Purchasing during peak demand times/Mass influx of demand limits supply ▪ Consistent (timeframe) and sporadic (quantity and type) demand ▪ Seeing and understanding the need to build relationships and generate a loyal customer base	Forecasting, Planning, and Preparation (Reactive vs. Proactive)	

To express the interconnectivity and complexity of these themes and dimensions, we present our findings and analysis using causal loop diagrams [79–81]. Senge [80] proposed this feedback-loop approach as a support to the challenges associated with human propensity to think in a linear fashion, given that seeing the entirety of the process is essential to understanding and solving a complex and dynamic problem, such as ours. In the context of our research, these diagrams help demonstrate how one challenge within the input supply chain interacts with others, leading to a vicious circle and exacerbating the poverty trap experienced by smallholders in Tanzania. Our analysis and all subsequent findings come directly from the data collected from the multiple participants that were interviewed during the course of our fieldwork.

3.5. Trustworthiness

Following Lincoln and Guba's [82] recommendations, we took several measures to enhance the trustworthiness of our research. First, interview and focus group guides were reviewed by experts from the FRI staff as well as two supply chain experts to ensure accuracy of wording for communication with farmers and for translation purposes, and to facilitate honest and forthcoming dialogue with participants. Second, the data obtained from over 100 smallholder participants via multiple focus groups was triangulated with information collected via interviews with other stakeholders. Finally, after the initial analysis, our data was double-coded by an external individual to the research team who reviewed both first-order and second-order codes, thus ensuring that our interpretation of the participant experience was coherent and supported by the data.

4. Findings

4.1. Mapping the Tanzanian Smallholder Rural Agro-Input Supply Chain

From our data, the regional input supply chain of the Meru District (Figure 2), although seemingly simple in terms of main actors and activities, is complex given the often-overlapping participation of multiple influencing actors. This complexity stems from the relationships that exist between these actors across varied exchanges of inputs, information, and money.

Primary inputs used by smallholders (seeds, fertilizers, and pesticides) are individually important elements in successful production, but they cannot be considered in isolation because certain inputs require the use of other inputs in order to be effective (i.e., hybrid seeds require more pesticide and inorganic fertilizer). We therefore consider them holistically as we examine product flow through the chain and henceforth refer to them collectively as inputs. These inputs are initially injected into the rural supply chain through large-scale suppliers who perform a variety of import, production, transformation (i.e., seed development and local seed production within Tanzania and blending of pesticides following import), and distribution functions, depending on the types of inputs each company engages with. They are typically located within city centers, often precluding direct access by smallholders due to the distance, time and cost associated with traveling from their villages and the inability to benefit from economies of scale. Inputs, accompanied by information on their proper use and application then flow to local (village-level) suppliers, who operate out of small one-room shops where conditions are not often favourable to storing perishable inputs such as seeds, nor large quantities of inventory. Inputs are then sold at retail quantities and prices to customers who are largely subsistence smallholder farmers, whose primary objective is to harvest enough crop to feed their families and, when possible, generate income by selling any surplus to support future input expenditures and other needs (i.e., school fees, medicine, other food, home improvements, and repairs):

> *"A small amount is sold to get some money for needs, for example, for school fees or for some other needs at home. [. . .] There is not a specific amount of food to keep, [we just try] to keep enough food to get to the next season."*
>
> —Smallholder farmer (Kikatiti village)

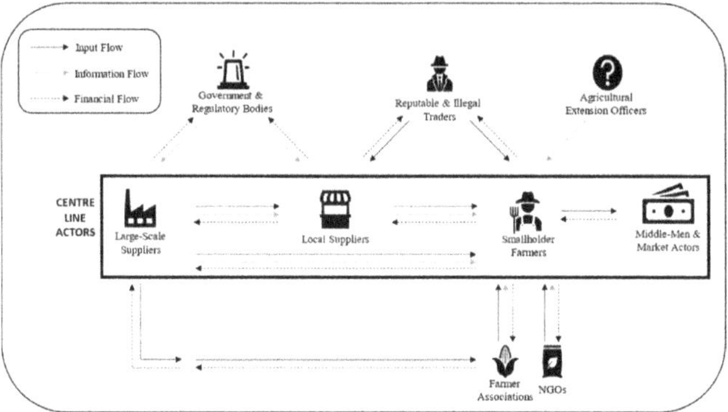

Figure 2. The Tanzanian Smallholder Rural Agro-Input Supply Chain.

In an effort to earn income, smallholders sell their surplus crops in local markets if accessibility permits, or they may sell to middle-men who then find market opportunities. Smallholders may also use farmer-saved seeds from their previous harvests, but this is not always possible due to challenges with germination and storage. The vast majority of interviewed smallholder farmers indicated that they had been farming all their lives and had remained on the same plot of family land, with exceptions in the case of marriage or a shift to smallholder farming from other employment. This finding highlights the limited land mobility of smallholder farmers who focus on agricultural crops without large livestock herds and who do not necessarily belong to pastoralist communities where they would have different opportunities to seek out supporting resources. Therefore, the smallholder farmers within our context have extremely limited land mobility which impacts their interaction with their supply chain and their ability to move towards areas that could support current or future resource requirements.

Beyond the primary actors (centre line) who move inputs and some information through the supply chain, there are multiple other actors who exert influence on it. These include regulatory bodies that conduct research, manage the training and licensing of large-scale and local suppliers, enforce regulations, ensure quality control, and certify inputs prior to sale by a large-scale supplier or provided by an NGO. Three regulatory bodies in Tanzania are (1) The Tanzania Official Seed Certification Institute (TOSCI), (2) the Tanzania Fertilizer Regulatory Authority (TFRA), and (3) the Tropical Pesticide Research Institute (TPRI). Traders may exert influence at the level of the local supplier and smallholder farmers to provide inputs, where illegal traders pose a significant problem within the system:

> *"Traders can pick grain and then dress them like our seeds, sometimes using the same packages we are using [...] we don't know where they are getting our packages. It's a problem for us, and a problem to the farmers."*
>
> —Large-Scale Supplier (Arusha)

Illegal traders sell uncertified or counterfeit inputs at enticingly low prices, ultimately diverting smallholders from purchasing quality inputs despite the oversight of regulatory boards. Although there are mechanisms in place to handle these actors if they are identified and caught, there is currently no way to completely discourage or stop this activity. AEOs in Tanzania are government officials specialized in various areas across a wide varieties of smallholder activities (including crops, livestock, and commercial). The role of the AEOs we interviewed was primarily centered around education and training of smallholder farmers. For example, they train farmers how to use fertilizer and pesticides correctly, and how to identify and use quality seeds. Finally, NGOs and Farmer Associations work to facilitate access to quality inputs or provide them directly, as well as offering other educational training activities.

With respect to the financial flow of the chain, the primary (and preferred) mechanism of exchange is cash, due to issues with obtaining, or using, credit or financing options at the level of both the local suppliers and the smallholders.

4.2. The Poverty Trap

As we delved into the implications of the relationships across the input supply chain for smallholder farmers, we identified multiple intricate and overlapping challenges that impede them from accessing quality inputs in an efficient manner, ultimately impacting their ability to generate enough income and pull themselves out of poverty: a phenomenon that we label the "poverty trap". We use a causal loop diagram to depict and define this trap, which is constructed on the basis of the aggregate dimensions we identified from our data and through our coding process. In the subsequent sections, we present the dynamics of the poverty trap step by step, adding a new loop (with each dimension, made up of multiple themes, representing a key challenge for smallholders), revealing the full complexity of connections between the various challenges experienced by smallholder farmers. Relationships between variables are displayed using directional arrows accompanied by "+" or "−" for positive or negative relationships.

4.2.1. Unequal Power Dynamics—The Heart of the Poverty Trap

> *"People with money [sellers], they have the power to speak to the government [...] we try our best, but it doesn't really work [...] because they have connections with people in power."*
>
> —Smallholder farmer (Kikatiti village)

The heart of the poverty trap is an unequal power dynamic between smallholder farmers and other actors (Figure 3), whereby more powerful actors are better positioned in negotiations, keeping smallholder farmers in a captive state and exposing them to risk. Although we focus on the variables and challenges that exist within the input component of the supply chain, it is necessary to highlight some market aspects to demonstrate the amplification of this unequal power dynamic and captive state. This simply means that the state of smallholder captivity within their capacity as a buyer of inputs is influenced and impacted by the captivity they experience as a supplier to the market. Pressure from the market squeezes smallholders into a situation where they need all the support they can receive to reduce cost and improve the quality of purchased inputs:

> *"We can produce good crops and take [them] to the market where we sell at whatever the market price is. [...] The inputs are expensive compared to the amount we get from selling at the market, which means that we don't have enough money to spend on the next process, the next season."*
>
> —Smallholder farmers (Karangai village)

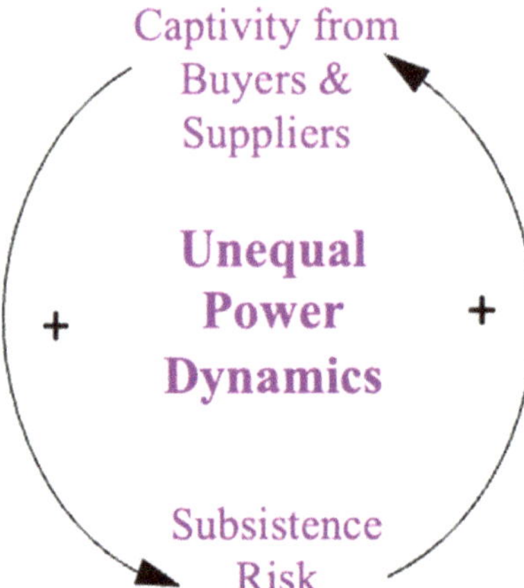

Figure 3. The Heart of the Poverty Trap.

Operating costs cannot be recovered, with one smallholder from Karangai village explaining that uncontrollable factors, such as climate change, are causing their costs to increase through the enhanced need for pesticides, for example, and market prices are not sufficient to account for this increase. With market saturation occurring at each harvest, selling price decreases and limits the smallholders' ability to accrue a decent income to cover necessary expenses. This results in a severe lack of on-hand capital, which forces smallholder farmers to sell what they can as soon as possible regardless of the price they may receive for their crops. As several smallholders noted, it is better to have some money than none. Many smallholders also lack the necessary resources to transport crops to market, forcing them to sell to middle-men at well below market value:

> *"If there was a specific market to sell [to], then we could try to find the transport to go there. But now, we have to sell with the middle-man."*
>
> —Smallholder farmer (Mbuguni village)

Further undermining the potential of profitable sales is the lack of appropriate facilities and equipment for storage or transformation of crops (e.g., turning tomatoes into tomato paste), which also forces them into making quick low-profit sales.

In addition to their market captivity, smallholders are held captive by local input suppliers. For example, benefit could be seen through using specific and innovative inputs (e.g., drought-resistant seeds), but these may take time to develop. It may take an extended period of time for these inputs to be available through local suppliers (based on their ability to acquire, stock, and sell) and if/when such products become available for smallholders to purchase, their high price renders them inaccessible to those with limited means, thus limiting smallholders' access to innovative and good-quality products. Smallholder purchasing habits are predictable, with inputs acquired routinely at the beginning of each planting season, which enables local suppliers to adjust prices for increased profit margins (sometimes regardless of government price controls), driving inputs further out of reach to smallholders. Smallholder farmers often lack the means to travel further than absolutely necessary and must therefore procure inputs from the closest supplier, who may not be in the same village. This effectively results in each local supplier holding a monopoly over their wide-spread customer base.

The captivity of smallholder farmers by more powerful actors at both ends of their operations generates subsistence risk through forcing smallholders into routinely purchasing the cheapest (and often lowest-quality) and most accessible inputs each season. This purchasing behaviour provides the potential for purchase of counterfeit or low-quality inputs. The quality of an input cannot truly be known until crops mature (or not), at which point the growing season may be over and any crop produced may not be sufficient; this results in lost time, money, and potential income, and smallholders risk being unable to feed their families:

> *"Even if I get the seeds, sometimes those seeds, [their] quality is not good, it's fake. [...] When I come to plant, I find out that the seeds are not original, it's fake, and they don't grow."*
>
> —Smallholder farmer (Kikatiti village)

While traders and local input suppliers may be guilty of providing substandard inputs, this may not necessarily be a conscious or malicious action. It is possible that they too are receiving substandard inputs from higher up the chain. In an attempt to mitigate the risk associated with quality issues, AEOs and local suppliers advise smallholders to read input packaging and look for certification and manufacturing labels. This is only minimally effective, even for those who are able to read, given the potential for package tampering by illegal traders. Smallholders are also encouraged to keep receipts and to keep some seeds in the original package as proof of purchase in case compensation or reimbursement might be possible. However, the process of returning inputs is time consuming and expensive for local suppliers because they are responsible to transport any returned seeds to the larger company for replacement. This process does not guarantee reimbursement for either, and it can negatively impact the local supplier if they have provided an initial reimbursement to the affected smallholder. There is thus little incentive for local suppliers to assist in compensating smallholders for defective seeds. Although one smallholder focus group noted that some NGOs will offer them compensation if inputs are of poor quality, the general consensus is that possibilities for reimbursement are extremely limited:

> *"Those people who sell us agricultural [inputs], like seeds, they don't care. They do their business. [...] They don't want to take back seeds."*
>
> —Smallholder farmer (Kikatiti village)

Further, if replacement seeds are offered to smallholders in place of financial reimbursement, these too may be of poor quality, and additional risk is incurred. Ultimately, any potential remedy comes too late, when the season is already lost.

This particular challenge may also be associated with issues of regulation and market surveillance that are certainly exacerbated by power imbalances, such as corruption, lack of state control, lack of accountability over supplier transactions, etc. Regardless, the risk to smallholders is significant and firmly rooted in a power imbalance whereby smallholders are unable to advocate for themselves. With each repetition of the cycle, smallholders are increasingly held captive, leaving them with less room to maneuver, negotiate, or take control over their input sourcing activities, thereby exposing them to a continual cycle of risk associated with uncertainty of crop sales and input acquisition, leading to subsistence risk. All other challenges outlined in the next sections connect to the power dimension through this risk variable, which in turn impacts the degree of captivity, which can then be followed through the remaining causal loops.

4.2.2. Access to Resources and Quality Inputs

> *"To be honest, quality has been a big problem for us, it left us poor and we have no solution on what to do. At the end of the day, it's wasting our time. We spend a lot of time to farm, to plant, [etcetera], but we don't meet our targets."*
>
> —Smallholder farmer (Kikatiti village)

Figure 4 shows the addition of challenges related to resource availability and quality inputs which are fostered by the unequal power dynamic and, in turn, reinforce the state of smallholder captivity and increased exposure to risk. The first variable accounts for the limited availability of resources for smallholder farmers, including credit/loans, capital, alternative payment mechanisms to cash, appropriate input storage facilities, and time, as well as the availability of quality inputs themselves. Any credit options offered by local suppliers are reserved for customers with whom they have a close relationship and bank loans are not an option for the bulk of smallholder farmers; 21% of smallholders who were asked about bank loans indicated that they had applied for a loan, with 18% having received one. One AEO noted that this low application and acceptance rate could be attributed to having little to no collateral and not being able to meet eligibility requirements. In an effort to assist with credit, AEOs encourage smallholders to create and participate in Village Community Banking Associations to collect savings and offer local loans to farmers, by farmers. However, the lack of start-up capital, the inability of individuals to contribute regularly and their lack of knowledge about how to coordinate such activities makes the activities difficult to implement. The absence of financing options limits both the quantity and quality of inputs that smallholders can purchase with their low cash-on-hand. Limited capital was noted as one of the primary challenges across all smallholder focus groups, sometimes to the point where inputs were not affordable at all:

> *"Overall, [...] the problem is lack of [capital] and the price is a little bit high, so we cannot afford at all."*
>
> —Smallholder farmer (Kwaugoro village)

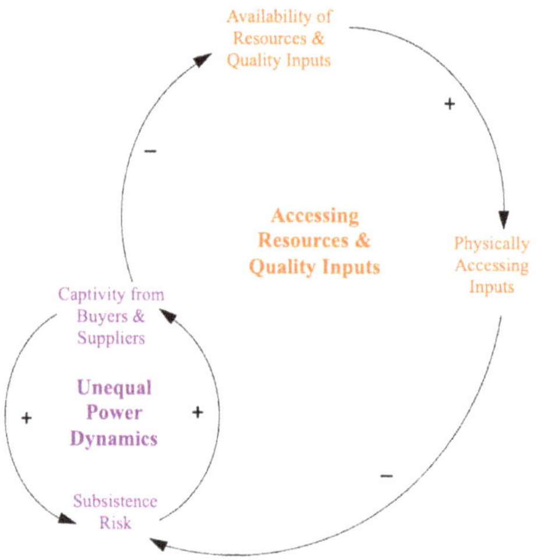

Figure 4. Exacerbation of The Poverty Trap (Stage 1).

To illustrate the earnings/cost ratio, on average, interviewed smallholders made 910,000 Tanzanian Shillings (TZS) (~$396 USD) from their previous season. To seed one acre of maize using the recommended quantity of 10 kg per acre, prices from one local supplier range from 35,000 TZS (~$15 USD) for the lowest quality product to 55,000 TZS (~$25 USD) for the better-quality product. As a more extreme example of seed prices, a 100 g bag of high-quality tomato seeds can cost as much as 360,000 TZS (~$154.50 USD), representing 40% of the average smallholder income. Thus, although quality inputs may be available on the market, they are not necessarily within reach of the smallholder farmers given their high cost.

Smallholders also lack the resources that would enable them to store inputs appropriately. Any storage facilities they may have access to are not capable of storing inputs for either short or extended periods of time because they do not allow for temperature and humidity control or for protection from pest predation. As such, smallholders are precluded from purchasing inputs ahead of the planting season and storing them, even if only for a short period. The final resource challenge is that of time; for example, in villages lacking irrigation systems, farmers spend much of their time collecting water, which reduces the time available to conduct higher-value activities, such as finding alternative suppliers and product sourcing.

The second variable highlights challenges of physically accessing inputs from suppliers. Smallholders participating in our study travel 2 to 30 km to reach local suppliers, over village roads that are often riddled with large rocks or potholes and can become further damaged by heavy rainfall. Even more challenging is traveling to the city (Arusha, the closest city to those villages within Meru District), which requires a 96 km round-trip to access alternative suppliers, further impeded by the need to access the main road via the same damaged village roads.

> *"Sometimes it's very difficult to get the seeds because our infrastructure is not that great. We have no transport most of the time. Sometimes the rain is heavy and there is flooding so you cannot move around to get the seeds."*
>
> —Smallholder farmer (Kikatiti village)

For farmers without personally-owned transportation, there is the added cost of travel via bus, car or motorcycle hire, if funds permit. In the worst case, they must walk, using a donkey or wheelbarrow to return home with their inputs. In turn, carrying capacity can be limited and the time required (which is already at a premium) increases; imagine an elderly smallholder walking for up to 30 km over rough roads with 50 kg (or more) of fertilizer and other necessary inputs using a wheelbarrow. Although some local suppliers offer assistance with delivery in exceptional circumstances, this is not usually the case. Limited and costly transportation options, minimal carrying capacity, poor road conditions, and the distance itself reduces the ability of smallholders to source inputs from other vendors:

> *"For example, if I live far away from the shop and there is another shop [closer by] with a little bit higher price, I have to buy, because I don't have the transport to go far away to get seeds or fertilizers [...]."*
>
> —Smallholder farmer (Kikatiti village)

Smallholders also indirectly bear the brunt of the distance and transportation costs incurred by local suppliers in transporting products from large-scale suppliers to their shops, which includes the cost of the transport itself, as well as the labour to load and unload trucks:

> *"For fertilizer, the government sets the price. For example, one bag of 50 kg, we have to sell it for 58,000 TZS, [and] buy it for 54,000 TZS. But the big problem is [...] [t]he cost to transport one bag from [town] to here is 1500 TZS. You also have labour costs to load/unload the truck. There is no profit in fertilizer."*
>
> —Local Supplier (Meru District)

On price-controlled fertilizer, where the government establishes the prices for all agrodealers across the input supply chain, these added transportation costs can prove detrimental to the local suppliers' bottom line and lead them to impose mark-ups on inputs beyond the allowable margins set by the government.

The issues identified in this loop present significant subsistence risk to smallholders. With the current cycle, they lack the necessary mechanisms in place to purchase, store, and transport quality inputs, as well as the means to pursue higher value activities, which challenges their ability to generate an adequate harvest to see them through the current season and into the next. This includes purchasing of new inputs, purchasing required items for the home (i.e., food that is not grown on their land) and providing for their families in general.

4.2.3. Access to Information and Support

> *"We never really had any education or instruction from agricultural officers. They don't really come and try to educate us on how to use seeds or to develop a proper routine of farming. We never really had an agricultural officer coming to help us."*
>
> —Smallholder farmer (Kikatiti village)

Challenges associated with accessing information and support are shown at Figure 5, and this perpetuates the challenges noted in the first two loops. The Meru District reported 102,134 smallholders across 94 villages at the time of our interviews, with only 34 village extension officers—well below the normal ratio of village extension officer to village at 1:1. For those AEOs whose purpose is to aid in agricultural training and education, being responsible for multiple villages over large geographical areas due to staffing shortages limits their ability to support smallholders, regardless of capability and motivation. This challenge was also noted by one large-scale supplier given that it relies on AEOs to provide essential product information to smallholders:

> *"Education is supposed to be delivered by extension [officers] and our country is so big. So the staff is not available. So farmers lack that education."*
>
> —Large-Scale Supplier (Arusha)

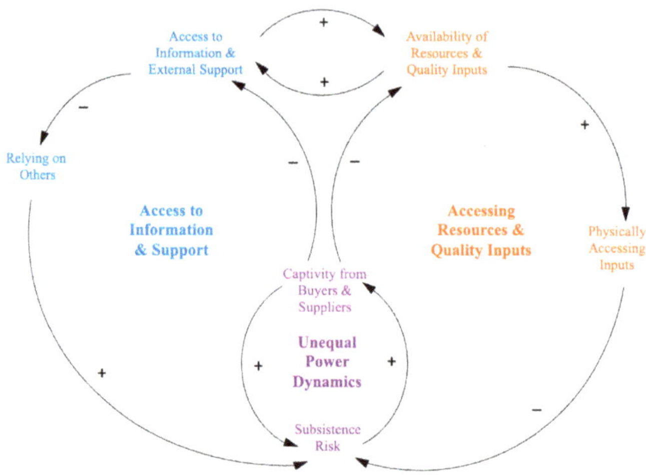

Figure 5. Exacerbation of The Poverty Trap (Stage 2).

Inadvertent, uninformed (or misinformed) use of inputs could result in substandard crops, which may be misconstrued as resulting from the purchase of poor-quality inputs rather than insufficient or flawed information. This could lead smallholder farmers into spending more money and time than necessary in the pursuit of quality inputs or indeed making a conscious decision to avoid using a particular input completely, which could hinder production. For example, one smallholder noted that many people are becoming sick due to a lack of knowledge on how to use agro-chemicals (pesticides). The possibility of negative health impacts due to improper (uninformed) use is often enough to deter someone from using a product which, in turn, causes them to miss the potential for higher crop yields, whereas with better information on how to use these hazardous inputs (e.g., using protective equipment), smallholders might be able to increase production. However, information alone may not provide the complete answer. As noted by one AEO, it is not always feasible for smallholders to obtain the recommended equipment due to its high cost:

> *"Up to 40% have knowledge [on pesticide use], but for most, the use of protective [equipment] is still a challenge. It's quite expensive. So most of them use ... overcoats they made themselves, gumboots. Instead of using gloves, they wear plastic bags, so that pesticides cannot come in contact with the skin."*
>
> —Agricultural Extension Officer

In an effort to mitigate information dissemination challenges, AEOs rely on local suppliers to provide critical product information to smallholders at the point of sale, particularly because local suppliers are required to attend mandatory training conducted through the various regulatory boards. However, as two local suppliers mentioned, attending these seminars can be costly in terms of money, time, and effort. Because smallholder farmers have little ability to access alternative suppliers, there is not necessarily an incentive for local suppliers to pay close attention during the seminars or take the time to provide information to customers, especially if it detracts from their sales. Furthermore, the sales person behind the counter may not have been the one to attend training events and cannot offer the correct information. AEOs acknowledge that this may not always be the most reliable mechanism by which to share information and educate smallholders:

> *"We as extension officers cannot reach all the farmers, so those [local suppliers] help us to give [smallholders] training. [However], some farmers, when you talk to them and ask about laws and regulations on how to use, maybe fertilizer and the precautions, they say they don't know. Because even if they go to the local market, the person who sells to them, knows nothing."*
>
> —Agricultural Extension Officer

While some conscientious local suppliers work to advise smallholder farmers on the benefits of quality despite the higher price, one local supplier suggested that farmers continue to choose the least costly inputs due to limited capital and a lack of trust in the information source.

AEOs also attempt to improve information dissemination through hosting village meetings. However, given the long distances to travel to attend these meetings, they are not accessible to all farmers, and some smallholders become aware of a meeting only after the fact. As a partial solution, AEOs encourage farmers to join groups where a representative may be sent to attend a meeting and pass on the information afterwards:

> *"We have also some farmer groups in the villages, so we advise farmers to make groups or be in their groups, to make easier work to train farmers."*
>
> —Agricultural Extension Officer

With the challenges of time and distance, it can still be difficult for farmer group representatives to connect with smallholders to pass on information. In some cases, a lack of capital and knowledge to organize and run such a group negates this as an option.

NGOs and other aid organizations are also actively involved in trying to close the information gap by providing education and training, particularly on input use. Many smallholders noted that these organizations have become the primary information source for them and are seen as more reliable and trustworthy than AEOs. This creates a secondary issue where trust is diminished, which widens the already existing gap between smallholders and AEOs and generates increased reliance on NGOs and other aid organizations for information.

The reliance of smallholder farmers on others is exacerbated by the difficulties they face in being able to directly access information themselves. Technological platforms, such as cellphones or computers, may offer smallholders the potential to find information (i.e., sourcing suppliers, comparing products, input use, etc.); however, while the majority of smallholder participants have cellphones, these are often of an older generation suitable only for communication, not for research or internet access. Furthermore, data networks that are fast enough to support this type of functionality are not available in the rural villages where smallholders live and work. Even if the technology was readily available, challenges with respect to literary rates would restrict those smallholders who are unable to read or write from accessing important information:

> *"To be honest, when it comes to check the quality of the product, it's really difficult for most of us, because most of us are not educated. So, it's really difficult, it's a big challenge for us because we don't know. Some of us, [we] don't know how to read and that's the problem."*
>
> —Smallholder farmer (Kwaugoro village)

The issues identified in this loop generate increased subsistence risk for smallholder farmers, compounding the risk already incurred through the first loop. With a lack of information and support on the right topics from the right providers coupled with challenges to access information on their own, smallholders become reliant on others for the limited information they have. Having some information is better than none, and it is not guaranteed that someone else will be able to offer much better. As such, they cannot take the chance to search for new sources of information because they may miss out on what already exists or may be subject to receiving worse information, or none at all. Missing out will mean that crops may be impacted in their yield or quality, thus posing a risk to supporting the family either through farm yields or any revenue that may be generated. From this loop, we see the challenges associated dependency/captivity coming from two loops.

4.2.4. Trade-Offs and Decision Making

> *"Quality seeds are there, but it's expensive. If you have good money, you can get quality seeds. Quality seeds are always there, [whenever] you get good money, you can get quality seeds. But [whenever] you have low money, you get low quality seeds."*
>
> —Smallholder farmer (Kikwe Village)

The final dimension to our causal loop diagram focused on trade-offs and decision-making processes and is added to Figure 6. This dimension speaks more to the business process of farming than the previous dimensions and depicts the reactive posture of smallholder farmers.

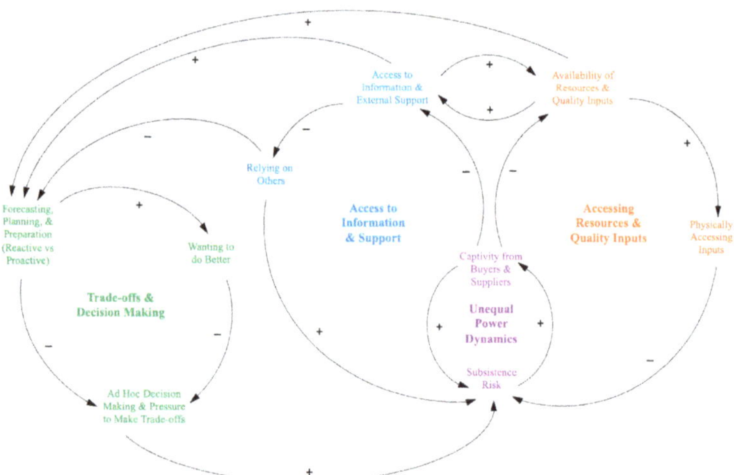

Figure 6. Trade-offs, Decision Making, and The Poverty Trap (Final Stage).

A lack of capital and credit implies that smallholders are not necessarily able to set money aside to purchase inputs in the following season, limiting their ability to forecast, plan, and prepare. This ultimately generates a significant amount of risk, with the imbalance tilted against the smallholders whose crops may not be sufficient for family needs, let alone for sale. To mitigate some of these challenges, smallholders will sometimes sell items, which were not originally intended for sale, to gain extra income. These items include any farmer-saved seeds from the previous season, livestock, milk, or eggs, which can offer a short-term solution to income uncertainty. This desperation-driven solution can have dramatic consequences for the family because such assets are generally very important to a household and are seldom intended for sale unless in exceptional circumstances:

"Sometimes we might even sell our livestock to get the money to go buy seeds or fertilizers, chemical fertilizers, then pesticides."

—Smallholder farmer (Mbuguni village)

A lack of information creates uncertainty about which inputs to purchase and when, which thus creates an environment in which options may not be fully weighed and where decisions are made in haste (to move to the next stages of planting). As just one example, smallholders, although advised against it by AEOs, wait for the rains to arrive before beginning their input sourcing, increasing the potential for reactive rather than proactive decision making due to an effort to procure and use inputs in the shortest time frame possible to minimize deterioration of input quality. As previously discussed, smallholders rely on their local suppliers to provide the necessary inputs, and if the desired inputs are not available, then the smallholder must make do with what is left on the shelves. As such, any proactive steps smallholders may be able to take in sourcing their inputs (e.g., planning for the quantity, price, or quality of required inputs, timing, etc.) are limited by the ability of the local supplier to support specific demand. Perhaps the most interesting theme we discovered is the desire to be better, where smallholder farmers know that there are better ways to conduct their activities. However, due to the accessibility issues they experience and not having the information necessary to coordinate activities amongst themselves, this desire to improve cannot be fulfilled. Several smallholder focus groups agreed that they would be willing to work more, spend more, or travel farther if they could be assured of a reputable supplier providing the quality inputs that would improve their crop yields. Nevertheless, the multiple challenges discussed in the previous sections impede their ability to do so:

"We don't really have capital to farm as much as we wanted. If we had capital and we organize ourselves as farmers, we can get our own agricultural equipment shops, we can get easier [access], so we don't have to go far away to get seeds and [other inputs]. Basically, we need capital to organize our farms and our [activities] so we can [improve] our farming industry."

—Smallholder farmer (Kikatiti village)

Although the smallholders we interviewed have made attempts to solve the issues they face, several groups noted that they have stopped asking questions and searching for solutions, given that they receive nothing back, and they no longer see this pursuit as beneficial to them. This again connects to subsistence risk, compounding what is already felt, where smallholders are even more unable to accept additional risk that could impact their ability to produce crops that are required to support their families. Although smallholders may sell other items to gain some money in order to purchase required items, this forced choice renders them more entrenched in poverty because now they have less than before in order to make the required purchases necessary to support their farming activities which in turn support their family.

4.2.5. The Importance of Subsistence Risk in the Poverty Trap

With the addition of each dimension to our causal loop, the subsistence risk variable is fed, which spurs the inequality of the power dynamic. Even faced with the high probability of earning consistently less income with each season, smallholders have no choice but to continue to buy their inputs and sell their crops in the same way as before because they cannot take on more risk, further entrenching them in a captive and impoverished state. This persistent cycle limits the smallholder's ability to take control of their input sourcing and improve their situation. This could become a vicious cycle, where the smallholder's continued inability to access the necessary resources that could shift the power dynamic makes their situation progressively untenable. We suggest that the power imbalance at the heart of this trap, and the variables that drive and sustain this imbalance, is the catalyst for poverty rather than its result. With the majority of these variables outside of the control of smallholder farmers, they do not have the capacity or ability to change the dynamic affecting their income (and thus their ability to make ends meet and provide for their families) season after season. They thus find themselves unable to escape a desperate situation. One smallholder farmer from Kikatiti village noted that farming is often the last resort for many. It is the last beacon of hope for them to support their families; when this 'last chance' entrenches them further into poverty through the multiple factors mentioned previously, the existential threat is very clear. It is this risk to everyday life that does not allow for improvements to be made and thus perpetuates the cycle of poverty across this demographic.

5. Ongoing Efforts in Overcoming Challenges

Our findings also identified some key stakeholders whose efforts are assisting smallholders in overcoming some of the challenges revealed in our analysis (Section 5.1). In Section 5.2, we offer some insights also based on our findings where additional support may be offered or required from these stakeholders (Figure 7).

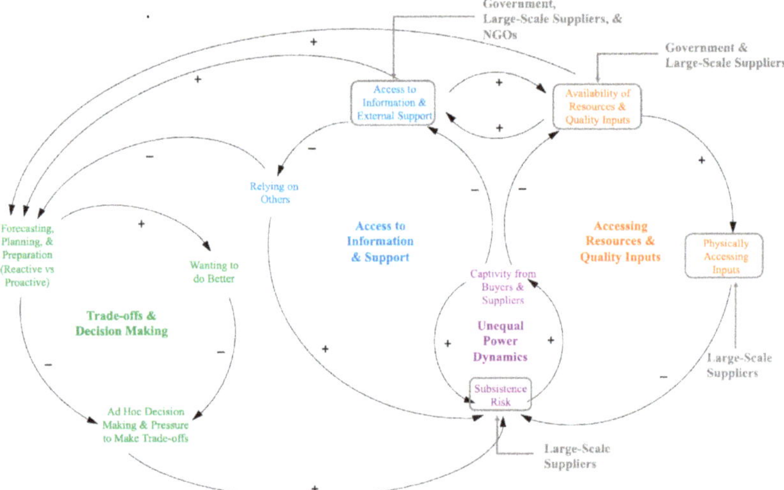

Figure 7. Stakeholders and Pressure Points in the Conceptual Framework.

5.1. Stakeholders and Targeted Variables

5.1.1. Government (AEOs)—Targeting Information, Support and Resources

We identify Agricultural Extension Officers, or AEOs, as having the potential to take on a more impactful role to stimulate positive change and to leverage those AEOs who may be more effective than others in developing relationships with local administrations and village leadership. The evidence in our case suggests that the effect of current AEO engagement at the community level has been disappointing. Our interactions with participants in our focus groups suggest that this is most likely attributable to a lack of trust and confidence in these officials due to smallholders receiving limited feedback when questions or concerns are identified. This presents an opportunity for AEOs to improve follow-up on smallholder queries which may then enable them to rebuild trust and the overall relationship with smallholders. This could subsequently improve their ability to provide smallholders with information that is heeded, thus offering meaningful benefit.

We also identify an opportunity for AEOs to target the Availability of Resources variable by contributing to policy development. These officials are well positioned to encourage government-backed initiatives that facilitate smallholder access to better resources (such as credit) and encourage banks or other institutions to provide loans for smallholders.

5.1.2. Large Scale Suppliers—Targeting Information, Support, Resources, Access and Risk Reduction

Meru Agro is a large-scale supplier located in the city centre of Arusha. They have made efforts to diversify their distribution network through their Lead Farmer initiative, which is based on a direct-to-smallholder model and offers an added benefit of reducing some of the challenges faced by smallholders, or at least for those smallholders taking part in the initiative:

> *"Meru Agro saw this as an opportunity to use lead farmers to [make available] inputs to smallholder farmers because, due to poor infrastructure in most rural areas it is not easy for them to access agro inputs. Most agro dealers are in town centers."*
>
> —Meru Agro Representative

Key to this initiative are lead farmers, who are selected based on their farming abilities or their informal leadership role in the community. Lead farmers receive training through Meru Agro on good agronomic practices focused on identifying and using quality inputs, details of Meru Agro products, and how similar assistance can be provided to other farmers. Lead farmers are also provided with certified/quality inputs from the company on a credit basis and free of delivery charges, with the loan repaid from the resulting sales. Meru Agro' representatives suggest that this approach has achieved some success over the last two years since its conception:

> *"It has increased farm yields due to use of quality inputs in integration with good agronomic practices [and] it has reduced the issue of fake inputs especially, seeds because now farmers are able to distinguish fake inputs from quality inputs."*
>
> —Meru Agro Representative

Through this initiative, Meru Agro provides an opportunity to shift the power balance in the system by bypassing the local suppliers and possibly illegal traders. Four variables of our causal loop are addressed through this direct-to-smallholder model, demonstrating that it is indeed possible to target multiple areas simultaneously. This initiative might contribute to redressing the risk and exploitive power dynamics felt by smallholders, while providing benefit to the large suppliers as well. This type of activity could benefit from additional study to further understand how it would contribute to improving smallholder livelihoods over the short and long term.

5.1.3. Nongovernment Organizations—Targeting Information

Farm Radio International (FRI) is one of the nonprofit organizations working to improve opportunities for smallholders in Tanzania. FRI works with radio broadcasting partners to share information with smallholders and to engage them through participatory communication practices. This approach to communication actively involves smallholders in the discussion, particularly through the formation of listening groups where they can listen to a broadcast together, discuss, and then have the opportunity to provide feedback to FRI, individually or as a group, to help identify ways by which information sharing can be improved or tailored to their needs. FRI provides an accessible platform even in remote areas (through household radios, or listening groups facilitated by FRI-provided radios), making information widely accessible with little to no financial penalty to the listener. In 2018/2019, FRI estimates that they, with their partners, reached 20 million people across rural Africa.

The volume of farmer-relevant information distributed though FRI's network of radio partners, combined with the important feedback loop of participatory communication, enables FRI to increase the knowledge of smallholders and therefore offers the opportunity for smallholders to have more agency throughout their activities. This can reduce the smallholder's reliance on others (and thus reduce their exposure to risk) and help rebalance the power dynamic in their interactions. Radio provides a convenient and inexpensive way to gain this information because farmers can listen to the radio concurrently with other activities. Learning 'over the air' means that literacy rates are not a concern, permitting the dissemination of the most relevant information to the most people. By enabling smallholders to further close the information gap, FRI helps shift the power dynamic in the farmers' favour, with potential to help overcome the poverty trap.

Table 4 provides a summary of the findings in this section.

Table 4. Current Initiatives.

Targeted Variable	Stakeholder	Current Initiatives	Impact on the Poverty Trap
Access to Information and Support	Government	• Increased AEO follow-up with smallholders • Trust and relationship development	Improve Information Sharing Decrease Reliance on Others Decrease Subsistence Risk
	Large-Scale Suppliers	• Improved agronomic practices • Training on input use • Training on quality identification	
	NGOs	• Knowledge development • Information dissemination	
Availability of Resources and Quality Inputs	Government	• Policy development • Government-backed initiatives for credit, loan, and banking access	Improve Information Sharing Decrease Reliance on Others Decrease Subsistence Risk
	Large-Scale Suppliers	• Minimizing impact of illegal traders • Quality assurance • Credit/loan options	
Physically Accessing Inputs	Large-Scale Suppliers	• Improved product availability • Delivery options—reducing transport related costs and distance travelled	Improved Ability to Physically Access Inputs Decrease Subsistence Risk
Subsistence Risk	Large-Scale Suppliers	• Collaborative efforts and risk sharing	Decrease Captivity from Buyers and Suppliers

5.2. *Opportunities for Sustainable Action*

In reference to our causal loop diagram and arising directly from our findings, we identify three primary variables where tangible and targeted improvement efforts can be made through coordinated stakeholder action with the goal of equalizing the power dynamic at the heart of the poverty trap, reversing the cycle, and offering an opportunity for sustainable activities to be implemented. The first variable is access to information and support such that smallholders could be better able to understand how to access and use inputs, and to better predict their needs and learn how to better forecast, plan, and prepare for their seasons. With continued guidance and support, farmers could feel more comfortable in using improved inputs and managing additional risk that comes from looking at alternative sources and types of supply. The second variable concerns access to the particular resource of capital, potentially through provision of credit, financing, and loan or micro-loan options because smallholders are not currently guaranteed an open-handed response from traditional banks, suppliers, or even Village Community Banking Associations (if applicable). Enabling private sector organizations through government involvement and underwriting the risk of non-repayment may be the only way to accomplish this. The final variable is that of quality inputs; however, we suggest that this cannot represent a viable long-lasting solution by itself, unless integrated with the two previously-mentioned areas. Providing quality inputs directly to the smallholder could offer improved crops and higher income, yet the yields remain dependent on uncontrollable factors such as weather and pests, so quality inputs in-and-of-themselves may not be sufficient to create enough long-term income growth to provide lasting benefit. Table 5 provides a summary of these findings.

Table 5. Key Variables to Leverage.

Variable to Leverage	Expected Impact for Smallholders
Access to Information and Support	• Improved understanding for smallholders on how to access and use inputs • Improved needs identification • Improved forecasting, planning, and preparation for planting • Increased understanding of and confidence in using improved inputs • Increased risk acceptance to source other suppliers
Access to Resources and Quality Inputs (Capital)	• Underwriting of risk from government and banking institutions
Access to Resources and Quality Inputs (Quality Inputs)	• Improved production and crop quality • Increased income from sales

When referencing Figure 7, the variable of subsistence risk is also identified. We assess that by targeting the aforementioned three pressure point variables, this will stimulate and promote the reduction of subsistence risk, where additional efforts by large-scale supplies may further promote risk reduction through activities noted in Section 5.1. We see in our causal loop diagram that all loops are connected through the subsistence risk variable, and it is this variable that drives the poverty trap. Thus, by collaborative activities targeting factors that will reduce this risk, we may begin to see a positive change in the cycle; as subsistence risk lowers, the impacts related to captivity will also decrease, with follow-on effects across the connected variables taking place, further reducing this risk, and so on. Table 5 provides a summary of these key variables to be leveraged.

With a more holistic and coordinated approach across these specific variables, we see the potential to move past unilateral solutions and to embark on a path towards long-lasting and sustainable improvement for smallholders underwritten by a comprehensive, system-based approach which respects the intricacies and interconnections of their complex input-sourcing processes. This approach may offer the potential for more opportunities for smallholders to begin to take control over their activities, reducing subsistence risk, thereby enabling smallholders to take on more risk to expand their operations and capabilities, and promoting a less captive state which may work to equalize the power dynamic in favour of smallholders with the goal of overcoming the poverty trap. We further suggest that it is through this type of effort that smallholders could be empowered to enhance the quality of their production and contribute even more significantly to the agricultural and economic development of their region.

To illustrate how this approach may be practically applied in support of ongoing innovative practices in agriculture, we offer some discussion of smallholder-run purchasing groups, which was one potential solution identified during our focus groups:

"That's a good idea, a very, very good idea to be organized all together and go buy all the agricultural equipment like seeds, fertilizer, and pesticides, that would be better. [...]"

—Smallholder farmer (Kwaugoro village)

"Buying as a group, is better because [...] the price is low and the quality is fantastic because the seller, they are too shy to cheat on the groups. [...]."

—Smallholder farmer (Kwaugoro village)

Smallholders were very enthusiastic about the prospect of purchasing their inputs as a group and could appreciate the potential value of this activity; however, they noted several impediments to success. This included lack of capital, the time required to travel to the large-scale suppliers who offer wholesale, an inability to coordinate the input type and quantity to accommodate multiple needs and individual priorities, and the lack of suitable storage facilities to store wholesale volumes:

"I think that is the best way to organize as a group. [...] But the challenge is everyone here has a different view of what they're going to plant for the season. [...] That's why we don't get organized and go buy the seeds together. Sometimes, you find out, one of us in the group doesn't have money and they cannot join the group. So, at the end of the day you just go and buy individually."

—Smallholder farmer (Mbuguni village)

Despite the impediments noted by smallholders, most of which we have previously identified as significant challenges through our causal loop analysis, we see the broad potential for this kind of initiative. We suggest that this could reduce reactive/ad hoc decision making through improved forecasting, planning, and preparation, minimize negative trade-offs, reduce overall costs, increase the chances of receiving good quality inputs, and provide the opportunity for smallholders to benefit from supplier-based incentives, such as delivery.

To achieve positive outcomes from this type of activity, we suggest that collective, coordinated stakeholder action would be desirable. For example, AEOs and NGOs could coordinate training for smallholders in how to organize and plan for input purchase and distribution within a large group with diverse needs (e.g., identifying overlapping and specific input needs), offer education to improve negotiating skills and understanding of contract management, and provide general oversight. Additionally, coordination between AEOs, NGOs, and various private industry stakeholders could promote accessibility to credit or financing options (either with banks or large-scale suppliers), offer wholesale quantities that are reasonable for a group of smallholders, provide enhanced delivery options, and enable innovative storage solutions (e.g., consignment-storage of inputs at the large-scale suppliers' temperature and humidity-controlled warehouses until required by the smallholder). Finally, we suggest that smallholders be actively engaged in the process, substantiated by their desire to improve the way in which they conduct their activities and willingness to make the necessary effort to find solutions, where the opportunity exists to do so at manageable risk:

"We tried to solve the problem and to manage the challenge[s] we have, but the problem is that we don't go far away. That's why when we heard that [you were coming] here, we heard the news last night and when we [got] up in the morning, straight away we came here. We think maybe you can solve our problems and to deal with the challenges. That's why we're here and we're glad to be here."

—Smallholder farmer (Kikatiti village)

Enabling a solution such as this may work towards reducing subsistence risk for smallholders as they may gain knowledge and confidence in applying alternative business practices that will reduce operating costs, allowing for greater overall revenue to be gained after sale. This may in turn offer a financial cushion for smallholders to be more comfortable in taking on greater risks to source new suppliers, improve processes, etc. This in turn may build rural resilience.

6. Discussions and Conclusions

The sustainability of food supply chains is the topic of an important conversation. Due to their vast numbers around the world, smallholder farmers are crucial actors to consider in discussions around food production sustainability [83]. In order for smallholders to continue contributing to global food networks on such a broad scale, we must find ways to identify, encourage and support sustainable agricultural practices, which we argue, begin with developing sustainable livelihoods for smallholders through reducing their susceptibility to the dynamics associated with poverty traps and reversing this vicious cycle.

Using the lens of supply chain management within this study, an interesting perspective is offered, where this field of study, although often focused on logistical components, also works to determine how value may be added across the chain. Within our context, smallholders may be able to benefit from the concept of value chain upgrading opportunities. However, several constraints exist, such as the availability of infrastructure and resources (including inputs, credit, and information), as well as the voids across governmental policies and regulations [84]. Smallholder farmers often find themselves within a horizontal supply chain, where relationships, collaboration, and social capital may enable increased purchasing power and improved input sourcing while offering a substitute for weaker institutions [84]. However, with trust at the centre of the incomplete, or informal, markets in which smallholders operate, this type of collaboration may be difficult if these important social networks cannot be established or maintained across the various stakeholders.

Through the opportunities to pursue value chain upgrading, we must be sure to incorporate sustainability principles that are applied as part of a continuous cycle requiring increased coordination. As part of a sustainable food value chain, main principles include performance measurement, understanding, and improvement within economic, social, and environmental dimensions [85]. Smallholders are indeed an important contributor to this chain, and their specific challenges must be contextualized within this.

With production capabilities being highly dependent on inputs and the associated limitations that exist for smallholders to access these inputs and other important resources such as credit, smallholders experience a significant risk to their ability to produce and participate in markets [86,87]. Our study's focus on inputs offers an opportunity to better understand the origins of this problem. In other words, our research presents a new perspective to understanding the challenges facing smallholder farmers in developing countries and the reasons why they remain entrenched in endemic poverty. Instead of focusing on market dynamics exclusively, we place greater emphasis on supply-side dynamics and relationships inherent to input sourcing to better understand the different constraints and challenges that may impact smallholders' ability to see increased value and sustainable livelihoods. We offer a holistic perspective to capture the variety of challenges faced by smallholders and their impact relative to each other, leading to a contextualized framework which presents a different perspective on the poverty trap—attributing it, at least in part, to the smallholder input supply chain. Our use of grounded theory enabled us to capture these challenges through the lens of the smallholder farmers and to map them by means of a causal loop diagram so as to highlight the accumulation, compounding, and complexity of the issues. As seen through our framework, multiple overlapping issues create severe challenges around input sourcing for smallholder farmers. This complexity and interconnectedness must be taken into account if any sustainable benefit is to occur to their benefit. Solution development and building rural resilience cannot occur in isolation and without contextualizing implementation as part of a holistic perspective; if challenges are approached as singular entities rather than as part of a larger whole with multiple cause and effects, we risk creating additional challenges not previously anticipated. We believe that our framework and our perspective have value beyond the specific setting studied, offering insight into how improved input sourcing might enable better standards of living across Sub-Saharan Africa and other regions where similar challenges exist.

Accountability and transparency between supply chain actors [60] is challenging for large, power actors within subsistence markets given gaps in regulations and capital [88]. We draw the conclusion that if it is challenging for those who can easily access information and money, and who possess the necessary connections to operate within these markets, then it becomes nearly impossible for those smallholders who possess much less power and access, making collaborative efforts imperative to achieving sustainability. Aligning with the UN Sustainable Development Goals on the elimination of poverty, the objective is to unwind the poverty trap in a sustainable fashion through balancing and redefining the power dynamic via subsistence risk reduction, giving smallholders a greater voice within their daily business transactions to improve their livelihoods and increase agricultural output, thus offering a greater impact to local and national economies. The end goal must be to shift the power dynamics so as to provide smallholders with a better platform from which they can sustainably address issues and improve practices across their supply chain.

A transdisciplinary and participatory training approach [89] is important to reducing exposure to risk over the long- and short-term, and in particular, subsistence risk as the primary area of concern for smallholders. This would also aid in building and maintaining stakeholder relationships, which can be the most difficult part of the collaborative process but is also the key to success [90], thus reducing the captivity of smallholders within the current environment. However, there is not always an incentive for stakeholders to engage in collaborative efforts. We see the NGO role as vitally important and being the primary entity to coordinate activities amongst stakeholders and aid in encouraging cooperation and collaboration. Furthermore, active smallholder engagement may dispel some of the damaging effects that could arise if collaboration is not managed and coordinated effectively, particularly once principal stakeholders leave and smallholders are left again to depend only on their own resources [91]. This may lead to a power void across stakeholders. By implementing an initiative such as small, localized, community-buying groups, this could offer one way for smallholders to reduce reliance on others and thus minimize the power distortions. A strong desire to improve current processes exists within the smallholder community and has been documented by Snyder et al. [92] who observed that smallholders were anything but slow to respond to more modern farming techniques. This mindset could be leveraged through more direct investment in smallholder farmers and supporting input markets to offer more impactful solutions. Given that conditions and characteristics will vary between smallholder agro-food value chains, there are opportunities for government stakeholders to explore and invest in alternative input and input provisioning options based on the particular environment (i.e., "organic vs. inorganic fertilizer") in close collaboration with the smallholders who are targeted by their extension services.

The poverty trap as it currently exists is a complex web that is challenging to unravel, and thus to correct. It will be even more difficult to reverse the downward spiral that we see today in such a way that it may offer a sustainable approach towards long-term poverty reduction.

7. Limitations and Future Research

It would be beneficial for future research to include a larger geographical area so as to compile data across various regions and therefore determine overlaps of challenges, or discover potential mechanisms by which resources may be combined to facilitate collaborative efforts. Additionally, throughout the theoretical sampling, we endeavoured to close as many gaps as possible; however, due to the macro perspective of this study and its complex nature, some peripheral questions remain open for further consideration. These questions include how credit can be provided to smallholders on a broad scale, understanding how collateral can be obtained, and how policy can be generated to improve information sharing across organizations.

Despite ongoing efforts to address various aspects of the poverty trap, it persists, so we must continue to ask why current interventions are thus far failing to break the cycle. What additional efforts are needed? Is this a failure of scaling? What can be done differently? Our research has aided in depicting the broad, complex, and interconnected nature of the poverty trap as it relates to the input supply chain, and we hope that this paper stimulates conversation and offers an avenue for further research within this area.

Although we have sought to expand the literature on poverty traps to highlight the contribution of the rural input supply chain, there is much opportunity for future research, such as expanding this study to follow the input supply chain out to the large-scale suppliers/importers (and past the rural side) to further investigate their constraints and challenges to add to the holistic picture of the complete input supply chain and to develop an output (i.e., market) component causal loop diagram to see how this intersects with and impacts the input component causal loop.

Author Contributions: Conceptualization, E.E. and M.-E.R.; Data curation, E.E.; Formal analysis, E.E.; Funding acquisition, M.-E.R.; Investigation, E.E.; Methodology, E.E., M.-E.R. and A.L.; Project administration, E.E.; Resources, M.-E.R. and A.L.; Supervision, M.-E.R. and A.L.; Validation, M.-E.R., A.L. and D.H.; Visualization, E.E., M.-E.R. and A.L.; Writing—original draft, E.E.; Writing—review & editing, M.-E.R. and A.L. All authors have read and agreed to the published version of the manuscript.

Funding: Funding was received by the Canadian Natural Sciences and Engineering Research Council [grant number 2014-03945] and by the Institute for Data Valorization.

Institutional Review Board Statement: The study was conducted in accordance with the Declaration of Helsinki, and approved by the Institutional Review Board (or Ethics Committee) of HEC Montréal (Project No. 2020-3722, Issued 8 October 2019).

Informed Consent Statement: Informed consent was obtained from all subjects involved in the study.

Data Availability Statement: Not applicable.

Acknowledgments: The authors extend their thanks to Farm Radio International for their support to this project, in particular, their enablement of data collection. We would also like to thank all those who offered encouragement and guidance over the course of this project, notably, Professor Stephan Vachon for his valuable comments and constructive feedback on our paper.

Conflicts of Interest: The authors declare no conflict of interest.

References

1. FAO. Policy Support and Governance—Sustainable Agribusiness and Food Value Chains. Available online: http://www.fao.org/policy-support/policy-themes/sustainable-agribusiness-food-value-chains/en (accessed on 9 March 2020).
2. World Bank. (25 February 2016). A Year in the Lives of Smallholder Farmers. Available online: https://www.worldbank.org/en/news/feature/2016/02/25/a-year-in-the-lives-of-smallholder-farming-families (accessed on 9 March 2020).
3. Ghatak, M. Theories of Poverty Traps and Anti-Poverty Policies. *World Bank Econ. Rev.* **2015**, *29*, 77–105. [CrossRef]
4. Olsson, L.; Opondo, M.; Tschakert, P.; Agrawal, A.; Eriksen, S.H.; Ma, S.; Perch, L.N.; Zakieldeen, S.A. Livelihoods and poverty. In *Climate Change 2014: Impacts, Adaptation, and Vulnerability. Part A: Global and Sectoral Aspects. Contribution of Working Group II to the Fifth Assessment Report of the Intergovernmental Panel on Climate Change*; Field, C.B., Barros, V.R., Dokken, D.J., Mach, K.J., Mastrandrea, M.D., Bilir, T.E., Chatterjee, M., Ebi, K.L., Estrada, Y.O., Genova, R.C., et al., Eds.; Cambridge University Press: Cambridge, UK; New York, NY, USA, 2014; pp. 793–832.
5. Bowles, S.; Durlauf, S.N.; Hoff, K. *Poverty Traps*; Princeton University Press: Princeton, NJ, USA, 2011.
6. Azariadis, C. The economics of poverty traps part one: Complete markets. *J. Econ. Growth* **1996**, *1*, 449–486. [CrossRef]
7. United Nations General Assembly. *Transforming Our World: The 2030 Agenda for Sustainable Development. A/RES/70/1*; United Nations: San Francisco, CA, USA, 2015.
8. Benson, T.; Mogues, T. Constraints in the fertilizer supply chain: Evidence for fertilizer policy development from three African countries. *Food Secur.* **2018**, *10*, 1479–1500. [CrossRef]
9. Mwinuka, L.; Mutabazi, K.D.; Graef, F.; Sieber, S.; Makindara, J.; Kimaro, A.; Götz, U. Simulated willingness of farmers to adopt fertilizer micro-dosing and rainwater harvesting technologies in semi-arid and sub-humid farming systems in Tanzania. *Food Secur.* **2017**, *9*, 1237–1253. [CrossRef]
10. Louw, A.; Jordaan, D. Supply chain risks and smallholder fresh produce farmers in the Gauteng province of South Africa. *South. Afr. Bus. Rev.* **2016**, *20*, 286–312. [CrossRef]

11. Kraay, A.; McKenzie, D. Do Poverty Traps Exist? Assessing the Evidence. *J. Econ. Perspect.* **2014**, *28*, 127–148. [CrossRef]
12. Giesbert, L.; Schindler, K. Assets, Shocks, and Poverty Traps in Rural Mozambique. *World Dev.* **2012**, *40*, 1594–1609. [CrossRef]
13. Mapila, M.A.; Njuki, J.; Delve, R.J.; Zingore, S.; Matibini, J. Determinants of fertiliser use by smallholder maize farmers in the chinyanja triangle in malawi, mozambique and zambia. *Agrekon* **2012**, *51*, 21–41. [CrossRef]
14. Markelova, H.; Mwangi, E. Collective Action for Smallholder Market Access: Evidence and Implications for Africa. *Rev. Policy Res.* **2010**, *27*, 621–640. [CrossRef]
15. Omiti, J.M.; Otieno, D.J.; Nyanamba, T.O.; McCullough, E.B. Factors influencing the intensity of market participation by smallholder farmers: A case study of rural and peri-urban areas of Kenya. *Afr. J. Agric. Resour. Econ.* **2009**, *3*, 57–82.
16. Barrett, C.B. Smallholder market participation: Concepts and evidence from eastern and southern Africa. *Food Policy* **2008**, *33*, 299–317. [CrossRef]
17. Chambers, R.; Conway, G.R. *Sustainable Rural Livelihoods: Practical Concepts for the 21st Century*; IDS Discussion Paper 296; IDS: Brighton, CO, USA, 1991.
18. Kalid, U.; Seuring, S.; Beske, P.; Land, A.; Ali Yawar, S.; Wagnar, R. Putting sustainable supply chain management into base of the pyramid research. *Supply Chain Manag. Int. J.* **2015**, *20*, 681–696. [CrossRef]
19. Sodhi, M.S.; Tang, C.S. Supply-Chain Research Opportunities with the Poor as Suppliers or Distributors in Developing Countries. *Prod. Oper. Manag.* **2014**, *23*, 1483–1494. [CrossRef]
20. Bhandari, P.B. Rural livelihood change? Household capital, community resources and livelihood transition. *J. Rural Stud.* **2013**, *32*, 126–136. [CrossRef]
21. Misiko, M.; Tittonell, P.; Giller, K.E.; Richards, P. Strengthening understanding and perceptions of mineral fertilizer use among smallholder farmers: Evidence from collective trials in western Kenya. *Agric. Hum. Values* **2011**, *28*, 27–38. [CrossRef]
22. Larson, B.A.; Frisvold, G.B. Fertilizers to support agriculture development in sub-Saharan Africa: What is needed and why. *Food Policy* **1996**, *21*, 509–525. [CrossRef]
23. Tura, M.; Aredo, D.; Tsegaye, W.; La Rovere, R.; Tesfahun, G.; Mwangi, W.; Mwabu, G. Adoption and continued use of improved maize seeds: Case study of Central Ethiopia. *Afr. J. Agric. Res.* **2010**, *5*, 2350–2358.
24. Kansiime, M.K.; Mastenbroek, A. Enhancing resilience of farmer seed system to climate-induced stresses: Insights from a case study in West Nile region, Uganda. *J. Rural. Stud.* **2016**, *47*, 220–230. [CrossRef]
25. Lotter, D. Facing food insecurity in Africa: Why, after 30 years of work in organic agriculture, I am promoting the use of synthetic fertilizers and herbicides in small-scale staple crop production. *Agric. Hum. Values* **2015**, *32*, 111–118. [CrossRef]
26. Gianessi, L.P. The Increasing Importance of Herbicides in Worldwide Crop Production. *Pest Manag. Sci.* **2013**, *69*, 1099–1105. [CrossRef]
27. Gramzow, A.; Batt, P.J.; Afari-Sefa, V.; Petrick, M.; Roothaert, R. Linking Smallholder Vegetable Producers to Markets—A Comparison of a Vegetable Producer Group and a Contract-Farming Arrangement in the Lushoto District of Tanzania. *J. Rural Stud.* **2018**, *63*, 168–179. [CrossRef]
28. Sheahan, M.; Barrett, C.B. Ten Striking Facts about Agricultural Input Use in Sub- Saharan Africa. *Food Policy* **2017**, *67*, 12–25. [CrossRef] [PubMed]
29. Snyder, K.A.; Miththapala, S.; Sommer, R.; Braslow, J. The yield gap: Closing the gap by widening the approach. *Exp. Agric.* **2017**, *53*, 445–459. [CrossRef]
30. Snyder, K.A.; Cullen, B. Implications of sustainable agricultural intensification for family farming in Africa: Anthropological perspectives. *Anthropol. Noteb.* **2014**, *20*, 9–29.
31. Rutsaert, P.; Chamberlin, J.; Oluoch, K.O.; Kitoto, V.O.; Donovan, J. The geography of agricultural input markets in rural Tanzania. *Food Secur.* **2021**, *13*, 1379–1391. [CrossRef]
32. Adjognon, S.G.; Liverpool-Tasie, L.S.O.; Reardon, T.A. Agricultural input credit in Sub-Saharan Africa: Telling myth from facts. *Food Policy* **2017**, *67*, 93–105. [CrossRef]
33. Madsen, S. Farm-level pathways to food security: Beyond missing markets and irrational peasants. *Agric. Hum. Values* **2022**, *39*, 135–150. [CrossRef]
34. Yami, M.; Van Asten, P. Policy Support for sustainable crop intensification in Eastern Africa. *J. Rural Stud.* **2017**, *55*, 216–226. [CrossRef]
35. Teklewold, H.; Kassie, M.; Shiferaw, B. Adoption of Multiple Sustainable Agricultural Practices in Rural Ethiopia. *J. Agric. Econ.* **2013**, *64*, 597–623. [CrossRef]
36. Hinderink, J.; Sterkenburg, J.J. Agricultural policy and the organization of production in Sub-Saharan Africa. *J. Rural Stud.* **1985**, *1*, 73–85. [CrossRef]
37. Kadjo, D.; Ricker-Gilbert, J.; Abdoulaye, T.; Shively, G.; Baco, M.N. Storage losses, liquidity constraints, and maize storage decisions in Benin. *Agric. Econ.* **2018**, *49*, 435–454. [CrossRef]
38. Raynolds, L.T. Fair Trade: Social regulation in global food markets. *J. Rural Stud.* **2012**, *28*, 276–287. [CrossRef]
39. Parrott, N.; Ssekyewa, C.; Makunike, C.; Ntambi, S. Organic Farming in Africa. In *World of Organic Agriculture*; IFOAM: Bonn, Germany, 2006; pp. 96–107.
40. Freidberg, S.; Goldstein, L. Alternative food in the global south: Reflections on a direct marketing initiative in Kenya. *J. Rural Stud.* **2011**, *27*, 24–34. [CrossRef]

41. Chamberlain, W.O.; Anseeuw, W. Contract farming as part of a multi-instrument inclusive business structure: A theoretical analysis. *Agrekon* **2017**, *56*, 158–172. [CrossRef]
42. Wang, H.H.; Wang, Y.; Delgado, M.S. The Transition to Modern Agriculture: Contract Farming in Developing Economies. *Am. J. Agric. Econ.* **2014**, *96*, 1257–1271. [CrossRef]
43. Barrett, C.B.; Bachke, M.E.; Bellemare, M.F.; Michelson, H.C.; Narayanan, S.; Walker, T.F. Smallholder Participation in Contract Farming: Comparative Evidence from Five Countries. *World Dev.* **2011**, *40*, 715–730. [CrossRef]
44. Oya, C. Contract Farming in Sub-Saharan Africa: A Survey of Approached, Debates, and Issues. *J. Agrar. Chang.* **2011**, *12*, 1–33. [CrossRef]
45. Bolwig, S.; Gibbon, P.; Jones, S. The Economics of Smallholder Organic Contract Farming in Tropical Africa. *World Dev.* **2009**, *37*, 1094–1104. [CrossRef]
46. Minten, B.; Randrianarison, L.; Swinnen, J.F.M. Global Retail Chains and Poor Farmers: Evidence from Madagascar. *World Dev.* **2009**, *37*, 1728–1741. [CrossRef]
47. Porter, G.; Phillips-Howard, K. Comparing contracts: An evaluation of contract farming schemes in Africa. *World Dev.* **1997**, *25*, 227–238. [CrossRef]
48. Bellemare, M.F.; Bloem, J.R. Does contract farming improve welfare? A review. *World Dev.* **2018**, *112*, 259–271. [CrossRef]
49. Meemken, E.M.; Bellemare, M.F. Smallholder farmers and contract farming in developing countries. *Appl. Econ.* **2020**, *117*, 259–264. [CrossRef] [PubMed]
50. Bellemare, M.F.; Novak, L. Contract Farming and Food Security. *Am. J. Agric. Econ.* **2016**, *99*, 357–378. [CrossRef]
51. Bellemare, M.F.; Lee, Y.N.; Novak, L. Contract farming as partial insurance. *World Dev.* **2021**, *140*, 105274. [CrossRef]
52. Ragasa, C.; Lambrecht, I.; Kufoalor, D.S. Limitations of Contract Farming as a Pro-poor Strategy: The Case of Maize Outgrower Schemes in Upper West Ghana. *World Dev.* **2018**, *102*, 30–56. [CrossRef]
53. Kabungo, A.M.; Jenkins, G.P. Contract farming risks: A quantitative assessment. *S. Afr. J. Econ. Manag. Sci.* **2016**, *19*, 35–52. [CrossRef]
54. Valkila, J. Fair Trade organic coffee production in Nicaragua—Sustainable development or a poverty trap? *Ecol. Econ.* **2009**, *68*, 3018–3025. [CrossRef]
55. Tsing, A. Supply Chains and the Human Condition. *Rethink. Marx.* **2009**, *21*, 148–176. [CrossRef]
56. Haider, L.J.; Boonstra, W.J.; Peterson, G.D.; Schlüter, M. Traps and Sustainable Development in Rural Areas: A Review. *World Dev.* **2018**, *101*, 311–321. [CrossRef]
57. Gehlich-Shillabeer, M. Poverty alleviation or poverty traps? Microcredits and vulnerability in Bangladesh. *Disaster Prev. Manag.* **2008**, *17*, 396–409. [CrossRef]
58. Barrett, C.B.; Garg, T.; McBride, L. Well-Being Dynamics and Poverty Traps. *Annu. Rev. Resour. Econ.* **2016**, *8*, 303–327. [CrossRef]
59. Mehta, K.; Semali, L.; Maretzki, A. The Primary of Trust in the Social Networks and Livelihoods of Women Agro-Entrepreneurs in Northern Tanzania. *Afr. J. Food Agric. Nutr. Dev.* **2011**, *11*, 5360–5372.
60. Mutonyi, S.; Beukel, K.; Hjortsø, C.N. Relational factors and performance of agrifood chains in Kenya. *Ind. Mark. Manag.* **2018**, *74*, 175–186. [CrossRef]
61. Bensaou, M. Portfolios of Buyer-Supplier Relationships. *Sloan Manag. Rev.* **1991**, *40*, 35–44.
62. Pfeffer, J.; Salancik, G. *The External Control of Organizations: A Resource Dependence Perspective*; Harper & Row: New York, NY, USA, 1978.
63. Kähkönen, A.K.; Lintukangas, K.; Hallikas, J. Buyer's dependence in value creating supplier relationships. *Supply Chain Manag. Int. J.* **2015**, *20*, 151–162. [CrossRef]
64. Caniëls, M.C.J.; Gelderman, C.J. Purchasing strategies in the Kraljic matrix—A power and dependence perspective. *J. Purch. Supply Manag.* **2005**, *11*, 141–155. [CrossRef]
65. Pfeffer, J.; Leong, A. Resource allocations in United Funds: Examination of power and dependence. *Soc. Forces* **1977**, *55*, 775–790. [CrossRef]
66. Cox, A. Power, value and supply chain management. *Supply Chain Manag. Int. J.* **1999**, *4*, 167–175. [CrossRef]
67. Hallén, L.; Johanson, J.; Seyed-Mohamed, N. Interfirm adaptation in business relationships. *J. Mark.* **1991**, *55*, 29–37. [CrossRef]
68. Chopra, S.; Sodhi, M.S. Managing Risk to Avoid Supply Chain Disruptions. *MIT Sloan Manag. Rev.* **2004**, *46*, 53–62.
69. Jaffee, S.; Siegel, P.; Andrews, C. *Rapid Agricultural Supply Chain Risk Assessment: A Conceptual Framework*; The World Bank: Washington, DC, USA, 2010.
70. Livingston, G.; Schonberger, S.; Delaney, S. Sub-Saharan Africa: The state of smallholders in agriculture. In Proceedings of the IFAD Conference on New Directions for Smallholder Agriculture International Fund for Agricultural Development, Rome, Italy, 24–25 January 2011.
71. CIA World Factbook. CIA World Factbook—Tanzania (2020). Available online: https://www.cia.gov/library/publications/the-world-factbook/geos/tz.html (accessed on 9 March 2020).
72. FAO. *Small Family Farms Country Factsheet: Tanzania*; FAO: Rome, Italy, 2018.
73. Arce, C.E.; Caballero, J. *Tanzania Agricultural Sector Risk Assessment*; World Bank Group Report Number 94883-TZ; World Bank: Washington, DC, USA, 2015; pp. 1–79.
74. Brinkhoff, T. Tanzania: Administrative Division, Regions and Districts. 2019. Available online: https://www.citypopulation.de/en/tanzania/admin/ (accessed on 1 June 2020).

75. Shosholoza. Ubicación de Tanzania en África. 2011. Available online: https://commons.wikimedia.org/wiki/File:Locator_map_of_Tanzania_in_Africa.svg (accessed on 1 June 2020).
76. Salkind, N.J. *Encyclopedia of Research Design (Vols. 1-0)*; Sage Publications, Inc.: Thousand Oaks, CA, USA, 2010.
77. Charmaz, K. *Constructing Grounded Theory: A Practical Guide Through Qualitative Analysis*; Sage Publications: London, UK, 2006.
78. Gioia, D.A.; Corley, K.G.; Hamilton, A.L. Seeking Qualitative Rigor in Inductive Research: Notes on the Gioia Methodology. *Organ. Res. Methods* **2013**, *16*, 15–31. [CrossRef]
79. Van Oorschot, K.E.; Akkermans, H.; Sengupta, K.; Van Wassenhove, L.N. Anatomy of a Decision Trap in Complex New Product Development Projects. *Acad. Manag. J.* **2013**, *58*, 285–307. [CrossRef]
80. Senge, P.M. *The Fifth Discipline: The Art and Practice of the Learning Organization*; Doubleday: New York, NY, USA, 1990.
81. Perlow, L.A.; Okhuysen, G.O.; Repenning, N.P. The Speed Trap: Exploring the Relationship Between Decision Making and Temporal Context. *Acad. Manag. J.* **2002**, *45*, 934–955.
82. Lincoln, Y.S.; Guba, E.G. Chapter 11—Establishing Trustworthiness. In *Naturalistic Inquiry*; Sage Publications Inc.: Beverly Hills, CA, USA, 1985; pp. 289–331.
83. Vadjunec, J.M.; Radel, C.; Turner, B.L., II. Introduction: The Continued Importance of Smallholders Today. *Land* **2016**, *5*, 34. [CrossRef]
84. Trienekens, J.H. Agricultural value chains in developing countries: A framework for analysis. *Int. Food Agribus. Manag. Rev.* **2011**, *14*, 51–82.
85. Neven, D. *Developing Sustainable Food Value Chains: Guiding Principles*; Food and Agriculture Organization of the United Nations (FAO): Rome, Italy, 2014. Available online: https://www.fao.org/3/i3953e/i3953e.pdf (accessed on 23 March 2022).
86. Poole, N. *Smallholder Agriculture and Market Participation*; Food and Agriculture Organization of the United Nations (FAO) and Practical Action Publishing: Rome, Italy, 2017.
87. Langyintuo, A. Smallholder farmers' access to inputs and finance in Africa. In *The Role of Smallholder Farms in Food and Nutrition Security*; Paloma, S.G.Y., Riesgo, L., Louhichi, K., Eds.; Springer: Cham, Switzerland, 2020; pp. 133–152.
88. Parmigiani, A.; Rivera-Santos, M. Sourcing for the Base of the Pyramid: Constructing Supply Chains to Address Voids in Subsistence Markets. *J. Oper. Manag.* **2015**, *33–34*, 60–70. [CrossRef]
89. Kerr, R.B.; Young, S.L.; Young, C.; Santoso, M.V.; Magalasi, M.; Entz, M.; Lupafya, E.; Dakishoni, L.; Morrone, V.; Wolfe, D.; et al. Farming for change: Developing a participatory curriculum on agroecology, nutrition, climate change and social equity in Malawi and Tanzania. *Agric. Hum. Values* **2019**, *36*, 549–566. [CrossRef]
90. Argenti, P.A. Collaborating with Activists: How Starbucks works with NGOs. *Calif. Manag. Rev.* **2004**, *41*, 91–116. [CrossRef]
91. Luwanda, M.C.; Stevens, J.B. Effects of Dysfunctional Stakeholder Collaboration on Performance of Land Reform Initiatives: Lessons from Community Based Rural Land Development Project in Malawi. *S. Afr. J. Agric. Ext.* **2015**, *43*, 122–134.
92. Snyder, K.A.; Sulle, E.; Massay, D.A.; Petro, A.; Qamara, P.; Brockington, D. "Modern" farming and the transformation of livelihoods in rural Tanzania. *Agric. Hum. Values* **2020**, *37*, 33–46. [CrossRef]

 *sustainability*

Article

Understanding Complex Relationships between Human Well-Being and Land Use Change in Mozambique Using a Multi-Scale Participatory Scenario Planning Process

Pedro Zorrilla-Miras [1,*], Estrella López-Moya [1], Marc J. Metzger [2], Genevieve Patenaude [2], Almeida Sitoe [3], Mansour Mahamane [4], Sá Nogueira Lisboa [3,5], James S. Paterson [2] and Elena López-Gunn [1]

1 I-Catalist S.L. C/Borni 20, Las Rozas, 28232 Madrid, Spain; elopezmoya@icatalist.eu (E.L.-M.); elopezgunn@icatalist.eu (E.L.-G.)
2 School of GeoSciences, The University of Edinburgh, Drummond Street, Edinburgh EH8 9XP, UK; marc.metzger@ed.ac.uk (M.J.M.); Genevieve.Patenaude@ed.ac.uk (G.P.); james.paterson@ed.ac.uk (J.S.P.)
3 Faculty of Agronomy and Forest Engineering, Eduardo Mondlane University, Maputo P.O. Box 257, Mozambique; almeidasitoe@gmail.com (A.S.); sanogueiralisboa@gmail.com (S.N.L.)
4 Centre Régional AGRHYMET, Université de Diffa, Niamey BP 11011, Niger; msourtchiani77@gmail.com
5 N'Lab, Nitidae, Agostinho Neto Avenue, Maputo P.O. Box 679, Mozambique
* Correspondence: pzorrilla-miras@icatalist.eu; Tel.: +34-622-644-620

Citation: Zorrilla-Miras, P.; López-Moya, E.; Metzger, M.J.; Patenaude, G.; Sitoe, A.; Mahamane, M.; Lisboa, S.N.; Paterson, J.S.; López-Gunn, E. Understanding Complex Relationships between Human Well-Being and Land Use Change in Mozambique Using a Multi-Scale Participatory Scenario Planning Process. *Sustainability* **2021**, *13*, 13030. https://doi.org/10.3390/su132313030

Academic Editor: John McDonagh

Received: 20 October 2021
Accepted: 19 November 2021
Published: 25 November 2021

Abstract: The path for bringing millions of people out of poverty in Africa is likely to coincide with important changes in land use and land cover (LULC). Envisioning the different possible pathways for agricultural, economic and social development, and their implications for changes in LULC, ecosystem services and society well-being, will improve policy-making. This paper presents a case that uses a multi-scale participatory scenario planning method to facilitate the understanding of the complex interactions between LULC change and the wellbeing of the rural population and their possible future evolution in Mozambique up to 2035. Key drivers of change were identified: the empowerment of civil society, the effective application of legislation and changes in rural technologies (e.g., information and communications technologies and renewable energy sources). Three scenarios were constructed: one characterized by the government promoting large investments; a second scenario characterized by the increase in local community power and public policies to promote small and medium enterprises; and a third, intermediate scenario. All three scenarios highlight qualitative large LULC changes, either driven by large companies or by small and medium scale farmers. The scenarios have different impact in wellbeing and equity, the first one implying a higher rural to urban area migration. The results also show that the effective application of the law can produce different results, from assuring large international investments to assuring the improvement of social services like education, health care and extension services. Successful application of these policies, both for biodiversity and ecosystem services protection, and for the social services needed to improve the well-being of the Mozambican rural population, will have to overcome significant barriers.

Keywords: multi-scale scenarios; participatory scenario planning; social-ecological system; poverty alleviation; land use change; nature's contributions to people; Mozambique

Research Highlights:

1. An increase in LULC change in Mozambique for 2035 is projected by all scenarios
2. The most important drivers are social empowerment and effective law application
3. Biggest differences stressed by policies that promote small or large-scale agriculture
4. The multi-scale approach reveals hidden differences in local economies
5. The participatory approach can be valuable to use in other least developed countries

1. Introduction

The successful transition towards a global society without extreme poverty by 2030 is one of the main objectives of the Sustainable Development Goals [1]. This transition should occur as part of a wider global shift to ensure human development occurs within planetary boundaries [2]). Changes in land use and land cover (LULC) are one of the principal drivers for the degradation of nature [3]. LULC change is currently causing the loss of 13 million hectares of forests every year and 12 million hectares of arable land are being degraded annually, affecting an estimated 1.5 billion people globally with a disproportionate amount (74%) hitting the poorest and most vulnerable [4]. The provision of many ecosystem services (ES) depend on land, so future LULC changes and land degradation will affect poor populations disproportionally, especially those compounded by a lack of alternatives [5,6].

Globally, an estimated 767 million people live in extreme poverty, with 42% living in the Sub-Saharan Africa region [7]. The majority (about 80%) live in rural areas and 64% work in agriculture. Poverty is a complex concept and there is not an international consensus on its definition; however, we understand poverty as the inability to meet minimum standards and functioning, such as access to clean drinking water and sanitation or having a minimum level of formal education [8]. Rural dwellers face greater difficulties in achieving some of those standards, such as, for example, reading, access to electricity or use of safely managed drinking water [7]. On the other hand, rural dwellers have access to a higher diversity of ecosystem services than urban residents.

Rural inhabitants rely on ecosystem services in many different ways: provisioning services for obtaining wood products for construction, tools and fuel; food like fruits, hunting animals, mushrooms, etc.; grass for livestock; regulating services such as a good quality water, land for agriculture, and climate services; and cultural services, like access to sacred places or areas for recreation [9,10]. In some cases, these ES help rural families as a coping strategy in critical situations or contribute to poverty alleviation [11,12]. Therefore, changes in wellbeing are expected to occur if the integrity of ecosystems providing essential ES for the poor are degraded [3,13,14].

Because of increasingly complex and unpredictable global circumstances, as well as a growing understanding of socio-ecological systems, land management and land use problems require solutions that acknowledge and manage uncertainty [15]. Reversing the trends of LULC degradation and promoting a sustainable poverty reduction strategy requires a deeper knowledge of the complex processes that drive and link LULC change and poverty reduction. There is a growing understanding that land use systems are dynamic and connected across scales [16,17], as well as the social, economic and environmental factors affecting poverty. Each region is affected differently by a wide range of drivers, which in turn are shaped by a globalised economy. Better knowledge and understanding of these multi-scale relationships will facilitate the provision of more realistic and holistic governance strategies [18,19]. For example, Butler et al. [20] found that stakeholders at higher levels proposed more transformative strategies than local stakeholders. Therefore, new insights into the relationships between LULC changes and the multiple dimensions of well-being require the examination of interrelations and interdependences between ecological and social systems across scales [21]. Radical transformations are needed for reaching long-term sustainable social-ecological systems, and creative and experimental approaches are needed to envision and conceptualize them [22,23].

The inclusion of stakeholders in research is a way to deal with uncertainty and bring science closer to the problems faced by managers and local communities [24,25]. This co-production approach is increasingly being used to produce useful research for decision-makers with complex, long-term and large-scale challenges [26–28]. One of the tools that is increasingly being used for co-creation, dealing with future uncertainties and creative thinking is participatory scenario planning, which has gained popularity in recent years [29–32]. The development of scenarios is a methodology frequently used to analyse the complex processes that drive changes in LULC [33–35]. Indeed, many researchers see the necessity in developing participatory scenarios processes across-scales to support transformational changes, to compare results and to better understand how cross-scale interaction will affect future societies [36–38]. In this study scenarios are considered descriptions of different plausible future situations that consider the uncertainty that exists beneath complex interactions of multiple factors. These scenarios are not predictions or forecasts, because they do not identify the most probable future [39,40]. In their simplest form, scenarios can be a vision for the future which can prepare individuals, communities and institutions for uncertainty and complexity through social learning [41], stimulating discussion and creative thinking [42].

Participatory scenarios have been used to study LULC changes in sub-Saharan Africa at regional [21,43–45] and local scales [19,45–49]; and there are several scenario exercises developed at multiple scales simultaneously [50–54]. Nevertheless, to the best of our knowledge, there is only one study that developed scenario exercises at multiple scales in sub-Saharan Africa [17]. There are also very few examples of the use of scenarios to address rural poverty alleviation in developing areas [20,55,56], and fewer still in Sub-Saharan Africa [45,47]. This work aims to contribute to this knowledge gap by creating national scenarios with subsets of linked regional scenarios. The resulting scenarios are not linked to global scenarios, which allows for new and independent trajectories to help increase ownership by the participants of the construction workshops.

We worked in Mozambique for a number of reasons: the research team has extensive experience working there; it is relatively politically stable (although in recent years it has seen large policy disruptions); it has high population growth; there are high poverty rates both in rural and urban areas; there are considerable economic development opportunities; small-scale farmers have a high direct relationship with their surrounding ecosystems; and it presents a dynamic mix of environmental risks and LULC changes (see Table 1 for figures about these aspects). Despite great economic development during the 1990s and 2000s, Mozambique still has one of the highest rates of poverty in the world [8]. This is probably due to a combination of the colonial history, two decades of civil war, recurring economic crises and climate related hazards. Agriculture is the main rural livelihood and represents 95% of rural employment and 20% of national GDP [57,58].

The overall objectives of the paper are: (1) to improve the understanding of the complex phenomena linking LULC change and wellbeing of small-scale farmers in sub-Saharan Africa; (2) to identify the driving forces of LULC change; (3) to contribute to the debate about possible futures in Mozambique; and (4) to illustrate the produced scenarios, so that they can be used as sub-Saharan African scenarios in future studies and policy settings.

To reach these objectives, we developed a multi-scale and participatory framework for rethinking land use in Mozambique, highlighting its relationship with human wellbeing. As a result, we developed one set of national scenarios and three sets of regional scenarios for the provinces of Gaza, Zambézia and Niassa (Figure 1). We developed the scenarios in 2015 to support building models connecting land use, ecosystem services and poverty alleviation. Previous studies revealed that biodiversity and poverty were inversely related, suggesting that poverty reduction would imply loss of biological diversity [59]. The relevance of the analysis of these scenarios is that these will help to improve the understanding of the complex phenomena linking LULC change and poverty reduction, which can support current policy decisions such as, for example, the preparation of the Nationally Determined Contributions under the United Nations Framework Convention on Climate Change, land use planning decisions or rural poverty programmes.

Table 1. Social and environmental facts and figures from Mozambique that help to frame the scenarios exercise and result.

Factor	Figures (Year)	Data Source
Economic Growth	GDP: (Mill USD) 2012: 11,608; 2018: 14,457 GDP per Capita (USD): 2012: 607; 2018: 490	[60,61]
State Budget	5637 Mill USD (2019)	[62]
Tourism	3.4% of GDP, 2.8% of total employment (2017)	[63]
Forest area	47% of the country has some kind of forest cover (34 million ha) (2016)	[64]
Agricultural technology	<10% of farms use improved seeds, <5% of farms use fertilizers and <10% of farms use animal traction (2015)	[65]

Table 1. *Cont.*

Factor	Figures (Year)					Data Source
Farming commercialization	Less than 20% of rural households sell their produce (TIA 2007).					[66]
Climate change	Increase of 1.5/3C in 2046–2065; Changes in raining patterns; Decrease 20% of agricultural production; Increase in extreme events. (2009)					[67]
Vulnerability to climate change	36% of farmers lost part of their crops because of droughts, 30% of farmers lost part of their crops because of floods					[65]
		Mozambique	Niassa	Zambézia	Gaza	
Population	2017	27.9 million	1.7 million	5.2 million	1.4 million	[68]
	Projected for 2035	43.8 million	3.2 million	8.1 million	1.6 million	[69]
Urban population	2014	21%	26%	21%	26%	[69]
	Projected for 2035	36%	26%	36%	26%	[69]
Population below 25 years old	2017	66%	69%	69%	64%	[69]
	Projected for 2035	60%	62%	65%	54%	[69]

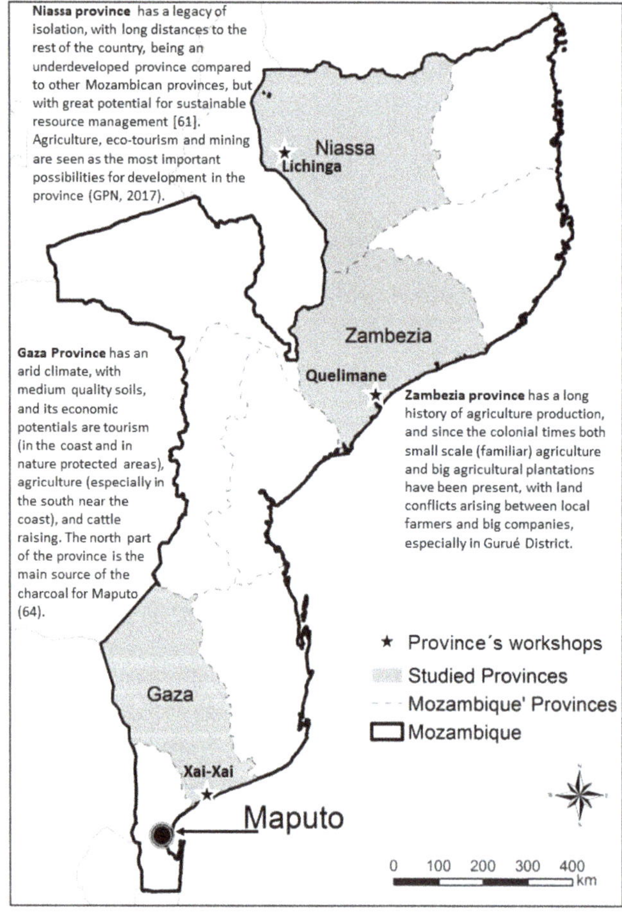

Figure 1. Map of Mozambique, with the provinces where the scenarios have been downscaled highlighted in grey colour. Sources of information: [67,70]. More relevant information about each province can be found in Supplementary S1.

2. Methods

2.1. Study Area

In order to have the most representative outputs possible, three contrasting provinces in the North, centre and the South of Mozambique were chosen. These provinces differ with contrasting land use transitions and pressures, as explained in Figure 1 and with more detail in Supplementary S1.

2.2. Multi-Scale Participatory Scenario Planning Approach

We followed a six-step approach (Figure 2), based on Metzger et al. [33]. A detailed description of scenario development methodology is included in Supplementary S2, and a summarized description below.

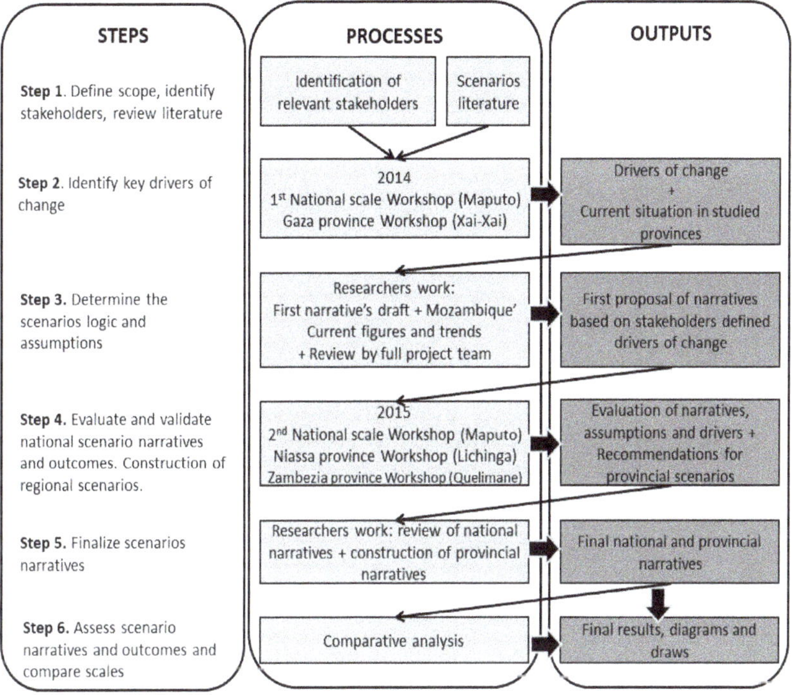

Figure 2. Methodological steps followed in the research, building on the methodology presented in Metzger et al. [33].

2.2.1. Step 1. Define Scope, Identify Stakeholders, Review Literature

The first stage was devoted to defining the goals and desired outcomes for scenario construction. This phase included reviewing previous scenario exercises in Mozambique and in neighbouring countries [55,71–73], and identifying a preliminary list of relevant drivers of change for land use, ecosystem services and human well-being.

A stakeholder analysis [74] identified the different types of Mozambican stakeholders from public, private, non-governmental organizations and academic institutions working in rural development, finance and management, environmental management, energy, agriculture, forestry, livestock and tourism.

2.2.2. Step 2. Identify Key Drivers of Change

A first round of workshops consisted of a workshop with stakeholders working with institutions at a national level (Maputo, August 2014: 23 participants); and one with stakeholders working at provincial and district levels working in the province of Gaza (Xai-Xai, August 2014: 14 participants; see Table 2 and Supplementary S3 for more details). The time-frame of the scenarios was defined through a consensus agreement.

Table 2. Number of participants representatives from each sector (government, private sector, NGOs and academia) in each of the five workshops developed.

Location Date Total Number of Participants	Number of Participants Representatives from Each Sector (Government, Private Sector, NGOs and Academia)
Maputo 12 August 2014 23 participants	5 participants from ministries (State Administration: Rural development, Agriculture: Environmental Management, Mineral Resources: Mines; 8 from provincial governments (Agriculture, Tourism, Planning and Finance, Rural energy market, and Environmental action), 2 participants from national NGOs; 3 from international NGOs; and 5 from Universities (Agriculture and forestry and Polytechnic).
Xai-Xai (Gaza province) 14 August 2014 14 participants	5 participants from the provincial government (Agriculture and Food Security; Forests and wild animals), 6 participants from district government (Economic Activities: the main governmental institution in the district), and 3 participants from local NGOs.
Lichinga (Niassa province) 4 August 2015 25 participatns	10 participants from the provincial government (Directorate of Agriculture, Directorate for Gender, Children and Social Action, Service of forests and wild animals, Niassa national reserve, Directorate of rural energy market, Directorate of Tourism), 2 from the district government (Economic Activities: the main governmental institution in the district), 1 from ecotourism, 1 from a private forest and wood processing company, 1 independent consultant, 4 from national-local NGOs, 2 from international NGOs, and 4 participants from universities (Education, Agriculture).
Maputo 12 August 2015 14 participants	3 participants from ministries (Wildlife Department; Directorate of Children, Adolescents and Family; Land, Environment and Rural Development), 1 from provincial government (Environmental Coordination), 3 from National NGOs, 1 from an international NGO, 1 from the National Institute of Disaster Management, 5 participants from universities (Agriculture and forestry, Socio-Economic Studies).
Quelimane (Zambezia province) 28 October 2015 21 participants	1 participant from the national government (REDD + Technical Unit of Ministry of Land, Environment and Rural Development), 6 participants from the provincial government (Directorate of Science and Technology, Directorate of Environmental Coordination, Services of livestock, Directorate of Land Environment and Development, Directorate of wood resources, Directorate of Economy and Finance), 2 from the district government (Services for Economic Activities, Services Planning and Infrastructure), 3 participants from wood and agricultural companies, 4 from national NGOs, 3 from the university (Marine and Coastal Sciences, University of Zambezia, Polytechnic University), 1 from the Gurué Agricultural and Livestock secondary Institute and 1 from the Mozambique Agricultural Research Institute–Zambézia.

During the two workshops, participants worked in groups to identify the main drivers of change affecting rural wellbeing, LULC and ecosystem services (check step 4 to see how we avoided missing important drivers because we developed this step at national level and in one province). Drivers of change were derived from participant's thoughts on what produced large transformations in society and the environment. Each group wrote down the key drivers of change structured into five categories: social, political, economic, technology and environment using the Ketso toolkit (© Ketso Ltd. 2018, Manchester, UK, www.ketso.com (accessed on 1 November 2020). The Ketso toolkit provides tags in the form of coloured leaves and branches of different sizes to display participants' contributions on felt mats. Each participant wrote at least one factor for each category, and then the groups continued adding drivers of change for the five categories. The general objective at this stage was to take into account as many drivers as possible to ensure that we did not miss any important driver of change.

Once the group had finished proposing drivers of change, they worked to identify the most important and most uncertain drivers. Each participant added 2 stickers to the most important drivers and 2 stickers to the less important ones and added the votes. Finally, after an internal discussion, the group agreed the 2 most important drivers (those causing the biggest changes from the current situation). Using the same method, they identified the 2 most uncertain drivers (those with the highest uncertainty about its future development) similar to the method used in Enfors et al. [46]. Those drivers are reflected in Table 3.

Table 3. The most important and uncertain drivers of change proposed by the participants during the first round of workshops, as proposed by each working group.

		Most Important Drivers of Change	Most Uncertain Drivers of Change
1st National Workshop	Group 1	• Empowerment of communities in the management of natural resources • Dissemination of laws and elaboration of land-use plans • Establishment of means for punishment (criminalization of adverse environmental impacts) + monitoring of the implementation of projects	• Recreational use of nature by inhabitants • Environmental protection
	Group 2	• Access to extension services • Effective application of legislation	• Fair prices • Reduction of gold digging (garimpo) and furtive hunting
	Group 3	• Decentralization and de-concentration with the participation of the civil society • Economic growth and development	• Effective decentralization • Economic development
Gaza province Workshop	Group1	• Social conflicts demanding development actions • More inclusive political decisions	• Improve rural technologies • Decreasing groundwater levels • An economy based on extractive industry
	Group 2	• Improvement of rural technologies • Improve environmental policies • Improving rural income	• Balances of payment (public deficit) • Importation/Exportation balance • Reforestation • Social protection
	Group 3	• Rural emigration • Effective application of legislation	• Improve employment opportunities • Erosion increase • The fragility of an effective application of legislation

2.2.3. Step 3. Determine Logic and Assumptions of the Scenarios: Post-Workshop Analysis and Construction of the First Version of Scenario Narratives

The scenarios had an exploratory goal [75] and a descriptive perspective [39]: the goal was to create a range of likely future alternative scenarios to examine plausible futures, therefore each scenario was not directed towards a single outcome (like normative scenarios do), but rather it was focused on exploring a range of plausible futures [40]. We (the research team using the inputs from workshops and from the literature) followed a combination of the "morphological" approach with the "intuitive logics methodology" [39]. The morphological approach visualizes all the possible interrelations between all potential factors, without prejudging the value of any of them [76,77]. Compared to the "two-axis" approach [78], the "morphological" approach is not restricted to two aspects (those that determine the axis), but rather incorporates a combination of different drivers, giving a similar importance to each of them, thus allowing the elaboration of complex and transparent scenarios. This approach can increase the relevance, coherence, plausibility, and transparency of the future alternative scenarios generated [77].

Following a morphological approach, we (1) clustered the drivers of change proposed by the participants in the five domains identified (following Metzger et al. [33]); and (2) searched for data supporting the current state of the different drivers. This meant we could propose future figures and combinations for these domains (see Supplementary S4: Table S1). From the full range of possible states, the drivers of change could take in the future, and the possible interrelations between them, we then followed the "intuitive logics methodology" [39]. For this, we (1) analysed the drivers and outcomes of the workshops to identify those considered most important by the participants in the workshops; (2) selected the key drivers, that were used to structure the future alternative scenarios; (3) based on the different possible future states of the key drivers, we selected a meaningful and coherent combination of drivers of change, and their possible future states, one for each of the different possible scenarios; (4) we wrote realistic and coherent descriptive narratives to explain the different outcomes of each scenario.

We decided to construct three scenarios; this number provides enough variability, but avoids adding too much complexity.

2.2.4. Step 4. Evaluate and Validate National Scenario Narratives and Outcomes. Construction of Regional Scenarios

With the same diversified range of stakeholders described in Step 1, we held a second and final set of workshops at national level in Maputo (October 2015, 14 participants) and at provincial level in Quelimane (Zambézia Province, October 2015, 21 participants) and Lichinga (Niassa Province, October 2015, 25 participants). The objectives of the national workshop were to evaluate the first version of the scenarios and to refine them to create a final version. Similarly, the provincial workshops evaluated the first version of the scenarios and produced a more refined version of provincial scenarios.

The three workshops followed the same process: participants were divided in five groups and each group worked with one thematic area (social, environmental, political, economic or technological). Each thematic group had to respond to the next set of 4 questions: plausibility of each scenario, whether one factor needed more attention, if any important driver or aspect was missing, and whether any driver should be taken out because of its low importance. Finally, each group explained to the other groups their results and a discussion followed.

2.2.5. Step 5. Finalize Scenarios Narratives

The narratives of the national scenarios were updated to incorporate the inputs from the second round of workshops (e.g., including a new driver of change or changing the assumptions in some of them). The scenarios were originally developed at the national scale, and then modified with the input of the provincial workshops to represent the contexts of each of the provinces. Comments from the participants about provincial-specific aspects were used to develop the provincial scenarios, adapting the national narratives to the provincial realities.

2.2.6. Step 6. Comparison of Provincial Scenario Narratives

Finally, the results for each province and for the national scenarios were compared and analysed. The main problems were identified and the agreed policies proposed in each province were compared, and differences and commonalities highlighted. The sequential elements of the narratives in each scenario were compared and finessed to ensure that policies were applied in different ways for each of them. The results of the comparison have been included in the discussion section.

2.3. Comparison of Scenarios Narratives with Actual Pathways

We compared the constructed scenario narratives with actual pathways in Mozambique since 2015, when the workshops and the narratives were built. The current situation has been matched to the most similar result of each scenario for the main drivers of change.

3. Results

3.1. Scope of Scenarios (Step 1)

Participants agreed a 20-year time-horizon for the development of the scenarios set for the year 2035, which aligns with the Mozambique National Development Strategy [60] developed for the period 2015–2035 (and operationalized through the government planning cycles and political agendas developed every five years).

3.2. Definition of Drivers of Change (Step 2)

The most important and uncertain drivers of change proposed by the participants in the first round of workshops (at National level and in Gaza province) are included in Table 3. The key drivers of change used to build the scenarios were: (a) Empowerment of communities and civil society, (b) Effective application of legislation, (c) Decentralization and a higher involvement of society in politics (e.g., more inclusive decision making), (d) Changes in rural technologies (both in communication and agriculture), (e) Economic growth and development, and (f) Migration. (Full data can be consulted in [79].

3.3. National Scenarios (Steps 3, 4 and 5)

The three scenarios represent different potential outcomes of LULC change and rural wellbeing in Mozambique for the year 2035 (Figures 3 and 4). Supplementary S4 contains full narratives of the scenarios.

Scenario A: Large private investments

Policies promote international and large-scale private sector as the main development motor; reduced local voice; low implementation of social and environmental policies; globalized approach to resource management. Capital investment increases Mozambique's GDP but equity in society declines, and most rural communities do not improve their livelihoods.

Scenario B: Small holder promotion

Social and environmental policies are successfully applied, in part because a demand from society: the proliferation of internet-based technologies, also in rural areas, increases the voice of local organizations and a more open and transparent governance model. Improved education and training in rural areas contributes to rural communities improving livelihoods.

Scenario C: Intermediate

Large private concessions increase, as well as education, with geographic differences; Internet-based technologies enable better democracy; Government economic resources increase from taxes from extractive projects. Some rural communities benefit from large commercial projects and others from improved social services; however, food security remains as the main concern for many communities.

Themes	Drivers	Scenario A	Scenario B	Scenario C
Politics	Social and environmental policy implementation	Weak	Strongly improved	Improved
Economy	Rural family incomes	Low	Improved	Some improved
	National GDP	High	Medium	High
Technology	Agricultural mechanization	In large projects	In local projects	Both
	Agricultural practices	High input	Conservation	Both
Environment	Forest cover	High decrease	Low decrease	Medium decrease
	Extraction industries	Intensive	Sustainable	Intensive
Society	Social services	Same as current	Improved	Improved
	Employment	Only in large projects	Med. in urban; High in rural	Slight increase

Figure 3. Designs, summaries, and key drivers of change describing respectively the national scale scenarios constructed via a participatory process for Mozambique in 2035. Data supporting the drivers of change can be consulted in Supplementary S4. Designs from the three scenarios by "Ross MacRae".

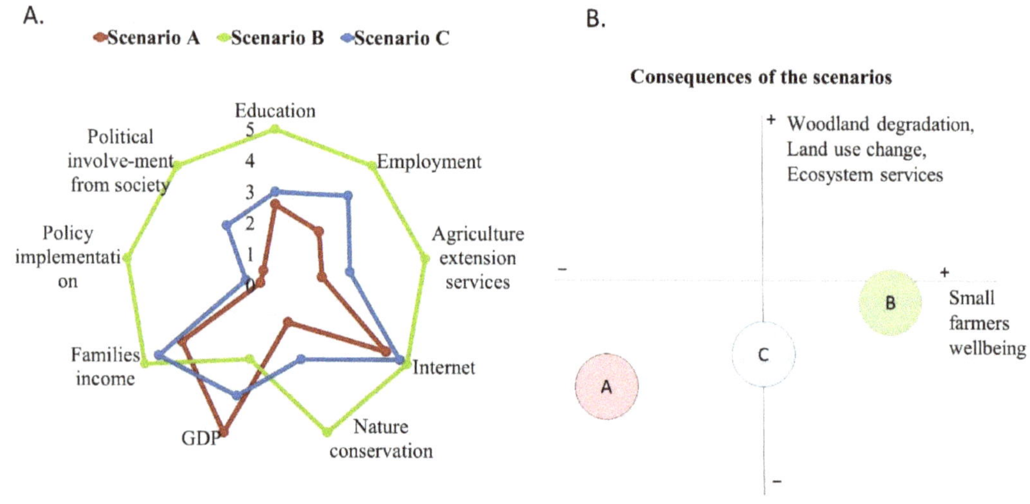

	Social		Technology		Environment	Economic		Policy and politics	
	Education	Employment	Agriculture extension services	Internet	Nature conservation	GDP	Families income	Policy implementation	Political involve-ment from society
Scenario A	2.6	2.3	1.6	4.2	1.3	5.0	3.6	0.5	0.6
Scenario B	5.0	5.0	5.0	5.0	5.0	2.5	5.0	5.0	5.0
Scenario C	3.0	3.8	2.5	4.7	2.5	3.8	4.4	1.0	2.5

Figure 4. (**A**) Spider gram representing qualitatively the main differences in each scenario to the different drivers of change. (**B**) Diagram representing the impacts of each scenario on small farmer well-being and the environment. Data supporting figures (**A**,**B**) can be consulted in Supplementary S4 (Tables S1 and S2).

3.3.1. Scenario A: Large Private Investments

Scenario A is characterized by public policies that promote international and large-scale private sector as the main development motor, accompanied by low implementation of social and environmental policy provisions. Scenario A presents also a reduced local voice (participation), and adopts a globalized approach to resource management. As a consequence, more of Mozambique's land is under private long-term leases and concessions by 2035. This includes agricultural and forested areas but also a significant increase in mined areas. The government favouring large foreign capital investment, together with an ineffective land use policy and an increase in technological advances, results in higher migration of rural populations to urban areas. Although capital investment considerably increases Mozambique's Gross Domestic Product (GDP), equity in society declines, and most rural communities do not improve their livelihoods (food security is their main concern). Implementation of state-led social and environmental policies is not effective due to lack of funding, e.g., public extension services continue to be scarce. Environmental quality also decreases in many ecosystems as a result of intensive land management. Mozambique's relations with its neighboring countries are improved through greater trading partnerships including China, many European Union countries, Brazil and India. Climate change adaptation and mitigation strategies are more reactive than proactive.

3.3.2. Scenario B: Small Holder Promotion

Local power is increased and public policies drive a development agenda focused on promotion and investment in small and medium enterprises. The proliferation of internet-based technologies, also in rural areas, increases the voice of local organizations, which pushes the government to increase public involvement in rural development and the improvement of public services. This scenario assumes there is also a real commitment from the government to improve education and training, and a more open and transparent governance approach. Social and environmental policies (e.g., education and training, health, water, extension services, and protected areas) are a priority for the government, partly due to societal demand in tandem with NGOs. Most rural communities improve their livelihoods: food sovereignty is achieved due to a sustainable and small-scale agriculture production, with a focus on extension services. Public support to communities results in sustainable forest management, which seeks to protect plant and animal diversity through harvest levels that respect ecosystem integrity. There are many areas for protected wildlife, and some are used for community-controlled eco-tourism. Mozambique welcomes international investments based on the requirement that companies respect local communities and share the development profits. Climate change adaptation strategies are strategically applied in small projects rather than in large programs.

3.3.3. Scenario C: Intermediate Scenario

This scenario presents a balance between a more globalized approach versus one with regional and local community empowerment in resource management. Large parts of Mozambique's land are in long-term private leases or concessions. However, an improvement in education, empowerment, and environmental stewardship allows some communities to self-organise and improve their well-being. Internet-based technologies enable better democracy and allow community empowerment to flourish in some areas of Mozambique, although the state still maintains a high control of resources and power. The economic government resources are higher because a greater percentage of income from taxes is levied on international extractive projects. This has special importance in some districts that have improved public services and community empowerment. Some rural communities benefit from large commercial projects, whereas other communities benefit from the improvement of social services. However, food security continues to be the main concern for the rest of the communities. There are several areas of protected wildlife, yet environmental quality decreases in many habitats and ecosystems as a result of intense use of resources. Climate change adaptation strategies are strategically applied in small projects rather than in large programmes, and there is an improvement in awareness raising, education and investment capacities.

The trends of land use land cover change under each future scenario are as follows:

1. Under Scenario A "large private investment", deforestation is driven by large companies that transform large parts of the country into agricultural land and achieve high rates of mining and timber extractions; urbanization is driven by rural migration to urban areas, in part due to the loss of land due to exploitation from private companies; woodland degradation is driven by the charcoal demand from the new urban inhabitants.
2. Under Scenario B "small holder promotion" rural families have a larger role and more power in decision-making resulting in agricultural land expanding into forests, more farm extension services and a growth of medium size farms. This scenario also assumes the government increases its capacity to enforce laws for the protection of natural areas. Nevertheless, the development of small scale farming and the increase in medium scale farming around the country also results in an increase in deforestation in non-protected areas.
3. Under Scenario C "Intermediate" both paths take place with similar intensity: agricultural expansion from small farmers, woodland degradation from charcoal demand and natural area degradation due to the impact of large investments in agriculture, mining and timber extraction. Nature protection is better achieved than in scenario A.

3.4. Province Scenarios (Step 6)

The downscaling of the national scenarios to the three provinces of Niassa, Zambézia and Gaza produced parallel scenarios with specifics in each of them (Table 4).

Table 4. Comparison of the scenarios in each Province.

	Niassa Province	Zambézia Province	Gaza Province
Introduction	The three scenarios imply a large expansion of infrastructures (roads and train connections) to facilitate agricultural expansion and transport of products. Interventions most voted by participants include: (a) the promotion of farmer's associations and (b) promotion of community natural resource management.	The interventions most voted by participants in the workshop include: (a) improving law compliance, (b) improving the transfer of agrarian technology to farmers to encourage conservation agriculture, (c) land use planning; (d) facilitating the process of acquiring land rights by farmers and the delimitation of communal areas.	The environmental consequences of charcoal production are a big concern, even if agriculture is the key economic activity. Proposed interventions: (a) to improve agricultural extension services and other agricultural services to increase farm mechanization and irrigation; (b) to improve the use of better seeds; (c) to promote alternative energy sources and improved charcoal stoves for urban consumers; and (d) to increase capacity building of rural communities.
Sub-Scenario A	Increase in the level of industrial activity, especially in mining operations. An increase in oil production in Niassa Lake opens a dispute in the Rovuma Basin between Malawi, Tanzania and Mozambique. Illegal timber operations grow due to the difficulties to obtain legal permits. PROSAVANA development project benefits especially big agricultural firms, producing the displacement of a large population to worse lands.	The government promotes large private agricultural schemes. Implementation of social policies is a challenge due to the number of private companies involved and a weak government capacity to enforce laws. Many farmers are moved from their lands, land conflicts increase between investors and smallholders, and a big part of the population migrates to other provinces, to cities, and to other countries.	The proposed interventions are not effectively applied by the government that is more focused on facilitating the implementation of large plantations, which occur mostly in the best agricultural land. Urban charcoal demand increases greatly, as a result of the great migration to urban centres, in part because of the problematic situation in the rural areas (see other provinces).
Sub-Scenario B	Big firms give up agriculture and forest plantations because of problems with bureaucracy. The government is successful in the promotion of irrigated agriculture, with big, medium and small infrastructures that allow farmers associations to increase their productions notably. An increase in tax revenues allows more access to credit by small farmers and more diversified job opportunities with most families improving their livelihood and wellbeing. Successful promotion of sustainable agriculture to small farmers, moving a high proportion of them out of poverty. PROSAVANA development project is directed to benefit small and medium scale farmers.	The promotion of conservation agriculture is successful (following an existing example by the NGO CLUSA). The government promotes small companies with public procurement procedures, like for small artisans and factories making pavements. In 2035 small companies are producing as a family sector. Improvement of access to IT in rural areas at accessible costs is achieved (as a combination of efforts from the government, NGOs, private companies and farmers). The use of solar panels increases (examples already exist in the province). Farmer movements obtain investments from the government and international bodies to improve water infrastructure in the Zambezi river, which increases agricultural production, especially for staple crops like rice.	The proposed interventions are applied successfully, since the government seeks to improve local rural capacities and nature protection. Urban charcoal demand remains constant, a result of low migration from rural areas to urban centres and an increase in the use of other types of energy, like renewable energies, that are promoted by the government and international organizations.
Sub-Scenario C	Reasonable investments in industrial development and more consciousness by taxpayers. PROSAVANA produces the displacement of some farmers, with others benefitting from the new infrastructure built, from private extension services and from a new variety of crops' value chain.	The problems from an informal style of doing things influence big investors, with a large part refusing to invest in Zambézia. Expansion of the Emergent Farmers' model: a greater proportion of land is controlled by medium size farmers (farming between 20 and 50 ha) increasing the production of horticulture and livestock and improving soil management.	Charcoal demand increases, but not as much as in Scenario A. Some interventions are successfully applied, especially those related to access to technologies, which facilitates communication for the population, who demand and achieve a substantially improved capacity for self-organization.

There are also differences in the main policies implemented, which reply to the different needs in each province. For example, in Niassa important policies imply the expansion of transportation infrastructures and the implementation of the PROSAVANA project, a large-scale national project directed towards the development of agriculture. In Zambezia, the improvement of extension services was detected as a crucial policy, together with improving the land tenure situation, which is critical nowadays because of the relations between large-scale investments and local farmers. In Mabalane, agriculture development needs improvements of infrastructure such as irrigation and improvements of the charcoal production.

3.5. Comparison of Scenarios Narratives with Actual Pathways

Since 2015, when the last workshops took place, Mozambique has taken a trajectory that aligns more with scenario A, but also partially with Scenario C (due to territorial differences) (Table 5). Since the scenarios were designed in 2015, important development landmarks have characterized the country. Particularly, the hidden debt crisis which arose in 2016, and slowed economic growth from about 7% to less than 4% and the Idai and Kenneth cyclones in 2019, which further reduced the economic growth to about 2%. It is expected that the COVID-19 pandemic will reduce the economic growth even more. This highlights the importance of aligning different interventions and following an integrative or systemic perspective, as individual initiatives can fail if they are not supported by complementary investments like infrastructure, social services or markets [46].

Table 5. Brief description of the Mozambican trajectory since 2015 and its correspondence with scenarios A, B and C.

Mozambican Trajectory Since 2015	Correspondence with Scenarios
The number of mine concessions has increased, increasing the power of large companies, and producing some negative effects on local farmers (i.e., conflicts in Cabo Delgado).	In line with Scenario A
The oil and gas sector took important measures with the final investments decision totaling more than 50 billion USD investment between 2017 and 2019 by multinational groups. This would have allowed an increase in social policies supporting small farmers. Nevertheless, due to the decrease in other sums (especially cuts in aid to governments by donors), social spending decreased.	In line with Scenario A (no increased support to small farmers).
Meanwhile, some areas benefitted from small and medium scale agricultural and forestry projects, e.g., a project funded by the World Bank (SUSTENTA project), in Zambézia and Cabo Delgado provinces, FAO projects and a Sweden supported project in Niassa province [80].	In line with Scenario C (territorial differences, with some areas benefitting and others not doing so).
Internet connections has not increased as in earlier periods. This is one of the drivers of change of the scenarios: Scenario B assumes there is a great increase in access to internet connections, which results in a higher civil society organization.	In line with Scenario A.

4. Discussion

4.1. Understand the Complex Processes That Link Land Use Change, Nature Degradation, and Poverty Alleviation

The resulting scenarios can be considered more adaptive than transformative [81], because they were designed more to look for interventions that could increase social and environmental sustainability under each different scenario than to look for potential sustainable futures. This has allowed a better understanding of the complex relationships between LULC change and local populations' well-being. Participants agreed that the main direct drivers of LULC change in Mozambique are the increase in agricultural land, urbanization, deforestation due to extractive activities like mining and timber production, and land degradation due to firewood and charcoal production (see full narratives in Supplementary S4). Under the different scenarios constructed, all these trends continue, but at different rates, patterns, and origins depending on the specific drivers and on the complex relationship between those drivers. Previous research indicated that a bottom-up or participatory resource governance would imply higher nature conservation results [82]. Although some participants of the presented research also agreed with this view, this was not totally agreed by all of them because bottom-up driven scenarios (in this case, Scenario B) can also head to high forest degradation and deforestation.

The scenarios show several complex interactions between drivers. For example, changes in rural technologies like small-scale solar panels and new communication technologies can allow the development of small-scale farms with a path that is less government-dependent compared to the current situation. Access to electricity with solar panels allows access to many other technologies (mobile phones, radio, refrigerators, TV, etc.) and changes in habits and social behaviour (e.g., enabling night-time study). The deployment of IT allows access to a wide knowledge repository stored in the web and facilitates communication and organization of civil society. This could be used both for increasing the demands pitched to the government and a better self-organization of communities. This is a key factor for development and poverty reduction. For example, stakeholders in Ethiopia considered that participatory forest management was useful to increase forest income in the long-term [47]. Nevertheless, the same study highlighted the difficulties faced by participatory initiatives due to weak accountability and growing inequalities or problems for controlling management decisions. More recent research also in Ethiopia recognized that participatory resource governance, and local agency would contribute to increasing natural capital and provide diverse harvests [82]. In our work, stakeholders proposed that the increase in societal leadership would allow civil society to push the government for improving social services. Scenario B is characterized by this process: an improved access to the internet in rural areas contributes to a significant improvement in education and extension services for small-scale farmers. This was identified as critical for a scenario of sustainable farmers' development in two research scenarios in Tanzania [46,48] and that previously referred to in Ethiopia [82].

Another example of the complex links concerns the effective application of the law. It was highlighted by most participants in the workshops with the common assertion that "Mozambique has good laws and plans that are not applied". Nevertheless, instigating effective application of the law and planning would result in important changes and very different futures depending on the government priorities. Participants commented that in the current situation, many large (mining, forestry and agriculture) companies are not investing in the country due partly to the poor application of the law, which decreases investors' confidence and certainty. Effective application of the rule of law will increase foreign investments in large projects, although in some cases this could increase conflicts with local inhabitants and decrease local well-being [83–85]. If these large projects were to occur, they would result in higher national tax revenues, which could potentially provide additional resources for improving social services. At the same time, more stringent application of the law could increase social services, and therefore increasing small-scale farmers' well-being.

Participants in the workshops highlighted the importance of peace as an important political factor in Mozambique. However, we did not include war as an option (i.e., a "shock event scenario"), because this would imply that any planned policy could not be implemented [71], and therefore it would have a small interest for policy makers. These events suggest that 'shock event' scenarios should be implemented in future projects, and methods like the OLDFAR algorithm [86] could be used in future scenario developments to achieve an optimally diverse set of scenarios.

4.2. Assessing the Multi-Scale Approach

Our scenarios are not embedded within global scenarios (c.f. [36]), but start from the analysis of national and regional driving forces. The method followed in our research allowed us to define the main driving forces for national and provincial levels simultaneously, involving national and local stakeholders. The process also allowed us to include regional and local perspectives in the national narratives enabling links between scenarios across different scales. Although the scenarios were first constructed at a national scale and then downscaled to regional areas, stakeholders from the provinces also proposed useful considerations for the national narratives. Their opinions were used both to evaluate the national narratives and to downscale them to the provincial scale. The involvement of stakeholders from the provinces provided knowledge of local realities, which was essential to root the scenarios in the reality of the country [82]. Following the framework set by Zurek and Henrichs [87] the scenarios presented should be classified as "consistent across scales": the regional scenarios share clear boundary conditions but each of them present different outcomes depending on the regional reality. The mixed method presented, by which participants contribute to the national scenarios and to configure provincial scenarios does limit the variability between provincial scenarios [17]. Nevertheless, the inclusion of local and provincial factors in the downscaling exercise allowed us to differentiate between provincial scenarios [88]. "Consistent across scales" scenarios are defined as useful for linking and comparing scenarios across regions [55], in line with Biggs et al. [44], who find the existence of loose links useful because they help maintain credibility and allow specific differences.

The multi-scale approach has highlighted the different consequences of scenarios in each scale and for each province. Across provinces, scenario A has the same impact: a decrease in natural areas, but due to different root causes (charcoal in Gaza, cropland in Zambezia and mines in Niassa). Another similarity across the three provinces is the vulnerability of small farmers, although they face different threats and opportunities in each province.

The differences between provinces arise from the different realities in each location: in Niassa mining activities and large forestry and agriculture companies have a contrasting effect on country revenues and ultimately GDP, but have direct negative effects on local farmers. In Niassa and Zambézia, the evolution of the ProSavana project is a good example of how quickly policies can change and curiously represents two contrasting elements of our scenario exercise. The original conception was to encourage industrial agriculture at large-scales but it has evolved to focus more on promoting small farmers due to public pressure. The real execution of this project still needs to happen. In Gaza, the third province, the northern districts will continue to be impacted by the urban demand of charcoal and could turn around their challenging situation due to droughts by an improvement in water infrastructure, which could also benefit the southern districts, that have higher farming and tourism potential. The participants in the provincial workshops evaluated the drivers and narratives proposed at national scales, and the final scenarios and narratives were adjusted to that evaluation. Therefore, our downscaling exercise allows us to describe more nuanced scenarios, with clear and precise examples of the consequences of the different plausible futures. In the presented case study, the downscaling has highlighted the importance of public policies to deal with external and internal driving forces.

The inclusion of quantified consequences of the three scenarios (e.g., future land use change in percentages, or future changes of regional poverty rates) was very challenging [89]. Reasons for this difficulty were the complex relations between drivers of change, the qualitative focus of the work developed, the lack of expertise from all participants concerning all aspects of the process (they were experts in just one specific sector), and the lack of time to work with rigorous quantitative results. Due to these circumstances, the results of quantifying land use change under each scenario contained contradictions and land use change rates much higher than the widely accepted ones. This learning implies specific time must be devoted to producing quantified results, and this process can benefit from an iterative process using modelling tools [89].

4.3. Policy Recommendations

The three scenarios show the difficulties the government face in improving the livelihoods of subsistence farmers. In order to achieve this, the government should promote agriculture extension services to tackle small-scale farmer productivity as well as increase farmlands size [90,91], and secure land tenure rights for small-holders [92]. However, the outputs of Scenario B suggest an increase in small-scale farmers' productivity and farm area could also increase deforestation and forest degradation [12]. In the face of a likely decrease in natural land cover in the next few decades, achieving effective protection of natural areas will be critical. Scenario A highlights the difficulty of controlling the actions of large companies, and the possible negative impact on small-scale farmers. Concentrating efforts to improve conservation of protected areas in the country [93] must involve local communities to ensure they also obtain benefits from nature protection [94]. Additional economic resources are needed for nature protection and management [14,95] and for impacted local populations. Part of those funds could be obtained from valuing the contributions ecosystems provide society, such as carbon sequestration or natural hazard regulation. Payments for Ecosystem Services schemes could be applied based on the lessons learned from the REDD+ programme. Additional revenues could also come from nature-tourism.

Previous participatory scenario construction processes have proved useful to support governance [89]. The participatory process presented in this paper and the project that developed it has influenced policy making in Mozambique already. Actions to influence policy include the publication of a policy brief and the presentation of the scenarios constructed in a final project conference. The scenarios have also been used to build Bayesian Belief Networks to model the consequences of different policies [12] and to produce maps representing land use change and ecosystem services distribution under each scenario in the province of Gaza [96]. Furthermore, participants in the workshops have been involved in the elaboration of public policies, and the research team has been consulted in those cases, resulting in conclusions from the research project being included in the policies. For example, Mozambique's forest policy has been reviewed and approved early in 2020 [97], calling for the development of a biomass energy policy as the basis for the promotion of sustainable charcoal production. The first NDC were submitted to the UNFCCC in 2018, recognizing that agriculture, forests and other LULC sectors have potential to contribute more than 80% of the greenhouse gas emission reduction [98], implying significant changes in the current dynamics of LULC. In addition, local measures have also been taken, such as the improvement of charcoal licensing and monitoring in Mabalane by the Gaza provincial Forest Service. Furthermore, several research activities have also been implemented to help improve understanding of land use dynamics and charcoal production (e.g., [99–101]).

5. Conclusions

The richer understanding and gains in context-specific knowledge on LULC and ecosystem services delivery for human well-being is particularly important in areas with populations of vulnerable small-scale farms. We have explored the interlinked consequences of drivers of change and how different these are when mediated through concrete decisions such as social and environmental policies and public extension services. These can have different context-specific and scale-dependent impacts on livelihoods, ecosystem services as well as LULC. We developed three plausible scenarios. Scenario A is characterized by the promotion of large-scale interventions. It highlighted large LULC changes from mining, agriculture and timber interventions resulting in increased migration from rural to urban areas but a negative impact on many rural livelihoods. Scenario B is centred on the promotion of small-scale farming as a result of societal pressures on the government. It would also produce large LULC changes due to the expansion of small and medium scale farmlands but would have the potential to bring about a more autonomous development and greater farmer empowerment. The capacity of the government for improving social services is necessary in this Scenario B. Higher participatory resource governance and local agency can trigger scenario B, which can be facilitated by new technologies like small scale renewable energy production and communication technologies. The middle road of Scenario C showed how large-scale projects linked with an effective application of the law can increase public resources, which can also be directed to promote the development of small-scale farmers and small private initiatives.

The results from this participatory scenario exercise aimed to support the exploration of options available to decision-makers in Mozambique. Several policy and decision-makers actively participated in the workshops, resulting in the creation of a policy brief. The co-creation process highlighted the value of the vision process and helped illustrate these complex and interlinked consequences by supporting the understanding and knowledge of trade-offs and synergies to improve future decisions.

Supplementary Materials: The following are available online at https://www.mdpi.com/article/10.3390/su132313030/s1, Supplementary S1: Description of the three studied provinces, Supplementary S2: Methodology followed in the development of the scenarios workshops, Supplementary S3: Detailed description of participants representatives from each sector, Supplementary S4: Full scenario narratives. References [102–115] are cited in the supplementary materials.

Author Contributions: Writing—original draft: P.Z.-M. and E.L.-M.; Writing—review & editing: P.Z.-M., M.J.M., G.P., A.S., S.N.L., J.S.P. and E.L.-G.; Conceptualization: P.Z.-M., M.J.M., M.M., G.P., J.S.P. and E.L.-M.; Data curation: P.Z.-M., M.M., A.S. and S.N.L.; Formal analysis: P.Z.-M., J.S.P., M.J.M. and E.L.-M.; Funding acquisition: M.J.M. and G.P.; Methodology: M.J.M., P.Z.-M., J.S.P. and M.M.; Validation: A.S., S.N.L. and E.L.-G. All authors have read and agreed to the published version of the manuscript.

Funding: This work was funded by the European Union's Horizon 2020 Research and Innovation Programme, under the Marie Sklodowska-Curie Grant, agreement number 798867 and by the ACES project (NE/K010395/1) from the Ecosystem Services for Poverty Alleviation (ESPA) programme (funded by the Department for International Development (DFID), the Economic and Social Research Council (ESRC) and the Natural Environment Research Council (NERC)).

Institutional Review Board Statement: Not applicable.

Informed Consent Statement: Informed consent was obtained from all subjects involved in the study.

Data Availability Statement: Data has been published in the referred data repository. Zorrilla-Miras, P., Matediane, J., Mahamane, M.; Nhantumbo, I.; Varela, R.; Metzger, M.J.; Patenaude, G. (2018). Scenarios of future land use change in Mozambique (2014 and 2015). NERC Environmental Information Data Centre. https://doi.org/10.5285/97c65c35-1db5-49d5-8ee0-ae5c7b699634 (accessed on 20 January 2019).

Acknowledgments: This paper is dedicated in memory of Raul Varela, who had a very active role in the development of the workshops that allowed the elaboration of this publication. Special thanks to Isilda Nhantumbo, who had a key role in the development of the conceptual elaboration of the scenarios and in the organization, facilitation and organization of the workshops. Special thanks to Maria Julieta Matediane for her relevant work in the stakeholder identification and in the logistical organization of the workshops, and to all the ACES team.

Conflicts of Interest: The authors declare no conflict of interest.

References

1. U.N. (United Nations). *Transforming Our World: The 2030 Agenda for Sustainable Development*; United Nations: New York, NY, USA, 2015.
2. O'Neill, D.W.; Fanning, A.L.; Steinberger, J.K. A Good Life for All within Planetary Boundaries. *Nat. Sustain.* **2018**, *1*, 88–95. [CrossRef]
3. Brondizio, E.S.; Settele, J.; Díaz, S.; Ngo, H.T. (Eds.) *Global Assessment Report on Biodiversity and Ecosystem Services of the Intergovernmental Science-Policy Platform on Biodiversity and Ecosystem Services*; IPBES Secretariat: Bonn, Germany, 2019; 1148p. [CrossRef]
4. Liniger, H.P.; Mekdaschi Studer, R.; Zander, U. *Making Sense of Research for Sustainable Land Management;* Centre for Development and Environment (CDE), University of Bern, Switzerland and Helmholtz-Centre for Environmental Research GmbH—UFZ: Leipzig, Germany, 2017.
5. Shackleton, C.M.; Shackleton, S.E.; Buiten, E.; Bird, N. The importance of dry woodlands and forests in rural livelihoods and poverty alleviation in South Africa. *For. Policy Econ.* **2007**, *9*, 558–577. [CrossRef]

6. Angelsen, A.; Jagger, P.; Babigumira, R.; Belcher, B.; Hogarth, N.J.; Bauch, S.; Börner, J.; Smith-Hall, C.; Wunder, S. Environmental income and rural livelihoods: A global-comparative analysis. *World Dev.* **2014**, *64*, S12–S28. [CrossRef] [PubMed]
7. U.N. (United Nations). *The Sustainable Development Goals 2017*; United Nations: New York, NY, USA, 2018.
8. Alkire, S.; Conceição, P.; Barham, A. *Global Multidimensional Poverty Index 2019; Illuminating Inequalities*; United Nations Development Programme and Oxford Poverty and Human Development Initiative: New York, NY, USA, 2019; p. 26.
9. Dewees, P.A.; Campbell, B.M.; Katerere, Y. Managing the Miombo Woodlands of Southern Africa: Policies, Incentives and Options for the Rural Poor. *J. Nat. Resour. Policy Res.* **2010**, *2*, 57–73. [CrossRef]
10. Fisher, M. Household Welfare and Forest Dependence in Southern Malawi. *Environ. Dev. Econ.* **2004**, *9*, 123–130. [CrossRef]
11. Arnold, J.E.M.; Ruiz-Pérez, M. Can Non-Timber Forest Products Match Tropical Forest Conservation and Development Objectives? *Ecol. Econ.* **2001**, *39*, 437–447. [CrossRef]
12. Zorrilla-Miras, P.; Mahamane, M.; Metzger, M.J.; Baumert, S.; Vollmer, F.; Luz, A.C.; Woollen, E.; Sitoe, A.A.; Patenaude, G.; Nhantumbo, I.; et al. Environmental conservation and social benefits of charcoal production in Mozambique. *Ecol. Econ.* **2018**, *144*, 100–111. [CrossRef]
13. Rodrigues, A.S.; Ewers, R.M.; Parry, L.; Souza, C.; Veríssimo, A.; Balmford, A. Boom-and-Bust Development Patterns across the Amazon Deforestation Frontier. *Science* **2009**, *324*, 1435–1437. [CrossRef]
14. Schreckenberg, K.; Poudyal, M.; Mace, G. *Ecosystem Services and Poverty Alleviation: Trade-Offs and Governance*; Taylor & Francis: London, UK, 2018; p. 352.
15. Folke, C.; Gunderson, L. Reconnecting to the Biosphere: A Social-Ecological Renaissance. *Ecol. Soc.* **2012**, *17*, 55. [CrossRef]
16. Holzhauer, S.; Brown, C.; Rounsevell, M. Modelling Dynamic Effects of Multi-Scale Institutions on Land Use Change. *Reg. Environ. Chang.* **2019**, *19*, 733–746. [CrossRef]
17. Kok, K.; Biggs, R.; Zurek, M. Methods for Developing Multiscale Participatory Scenarios: Insights from Southern Africa and Europe. *Ecol. Soc.* **2007**, *12*, 8. [CrossRef]
18. Brown, C.; Holzhauer, S.; Rounsevell, M. Land Managers' Behaviours Modulate Pathways to Visions of Future Land Systems. *Reg. Environ. Chang.* **2018**, *18*, 831–845. [CrossRef]
19. Cinderby, S.; Bruin, A.; Barron, J. Participatory geographic information systems for agricultural water management scenario development: A Tanzanian case study. *Phys. Chem. Earth Parts A/B/C* **2011**, *36*, 1093–1102. [CrossRef]
20. Butler, J.R.A.; Bohensky, E.L.; Suadnya, W.; Yanuartati, Y.; Handayani, T.; Habibi, P.; Park, S.E. Scenario Planning to Leap-Frog the Sustainable Development Goals: An Adaptation Pathways Approach. *Clim. Risk Manag.* **2016**, *12*, 83–99. [CrossRef]
21. Scholes, R.J.; Reyers, B.; Biggs, R.; Spierenburg, M.J.; Duriappah, A. Multi-Scale and Cross-Scale Assessments of Social–Ecological Systems and Their Ecosystem Services. *Curr. Opin. Environ. Sustain.* **2013**, *5*, 16–25. [CrossRef]
22. Bennett, E.M.; Solan, M.; Biggs, R.; McPhearson, T.; Norström, A.V. Bright Spots: Seeds of a Good Anthropocene. *Front. Ecol. Environ.* **2016**, *14*, 441–448. [CrossRef]
23. Pereira, L.M.; Hichert, T.; Hamann, M. Using Futures Methods to Create Transformative Spaces: Visions of a Good Anthropocene in Southern Africa. *Ecol. Soc.* **2018**, *23*, 19. [CrossRef]
24. De Vente, J.; Reed, M.S.; Stringer, L.C.; Valente, S.; Newig, J. How Does the Context and Design of Participatory Decision Making Processes Affect Their Outcomes? Evidence from Sustainable Land Management in Global Drylands. *Ecol. Soc.* **2016**, *21*, 24. [CrossRef]
25. Reed, M.S. Stakeholder Participation for Environmental Management: A Literature Review. *Biol. Conserv.* **2008**, *141*, 2417–2431. [CrossRef]
26. Beier, P.; Hansen, L.J.; Helbrecht, L.; Behar, D. A How-to Guide for Coproduction of Actionable Science. *Conserv. Lett.* **2017**, *10*, 288–296. [CrossRef]
27. Norström, A.V.; Cvitanovic, C.; Löf, M.F.; West, S.; Wyborn, C.; Balvanera, P.; Campbell, B.M. Principles for Knowledge Co-Production in Sustainability Research. *Nat. Sustain.* **2020**, *3*, 182–190. [CrossRef]
28. Wyborn, C.; Datta, A.; Montana, J.; Ryan, M.; Leith, P.; Chaffin, B.; Miller, C.; Van Kerkhoff, L. Co-Producing Sustainability: Reordering the Governance of Science, Policy, and Practice. *Annu. Rev. Environ. Resour.* **2019**, *44*, 319–346. [CrossRef]
29. Kok, K.; Pedde, S.; Gramberger, M.; Harrison, P.A.; Holman, I.P. New European Socio-Economic Scenarios for Climate Change Research: Operationalising Concepts to Extend the Shared Socio-Economic Pathways. *Reg. Environ. Chang.* **2019**, *19*, 643–654. [CrossRef]
30. Oliveira, A.S.; de Barros, M.D.; de Carvalho Pereira, F.A.; Gomes, C.F.S.; Da Costa, H.G. Prospective scenarios: A literature review on the Scopus database. *Futures* **2018**, *100*, 20–33. [CrossRef]
31. Oteros-Rozas, E.; Martín-López, B.; Thyresson, M. Participatory Scenario Planning in Place-Based Social-Ecological Research: Insights and Experiences from 23 Case Studies. *Ecol. Soc.* **2015**, *20*, 32. [CrossRef]
32. Rutting, L.; Vervoort, J.; Mees, H.; Driessen, P. Participatory Scenario Planning and Framing of Social-Ecological Systems: An Analysis of Policy Formulation Processes in Rwanda and Tanzania. *Ecol. Soc.* **2021**, *26*, 20. [CrossRef]
33. Metzger, M.J.; Rounsevell, M.D.; Hardiman, P.S. How Personal Judgment Influences Scenario Development: An Example for Future Rural Development in Europe. *Ecol. Soc.* **2010**, *15*, 5. [CrossRef]
34. Verburg, P.H.; Tabeau, A.; Hatna, E. Assessing spatial uncertainties of land allocation using a scenario approach and sensitivity analysis: A study for land use in Europe. *J. Environ. Manag.* **2013**, *127*, S132–S144. [CrossRef]

35. Rounsevell, M.D.A.; Pedroli, B.; Erb, K.-H.; Gramberger, M.; Busck, A.G.; Haberl, H.; Kristensen, S.; Kuemmerle, T.; Lavorel, S.; Lindner, M.; et al. Challenges for land system science. *Land Use Policy* **2012**, *29*, 899–910. [CrossRef]
36. Nilsson, A.E.; Bay-Larsen, I.; Watt, L.M. Towards Extended Shared Socioeconomic Pathways: A Combined Participatory Bottom-up and Top-down Methodology with Results from the Barents Region. *Glob. Environ. Chang.* **2017**, *45*, 124–132. [CrossRef]
37. Totin, E.; Butler, J.R.; Sidibé, A.; Partey, S.; Thornton, P.K.; Tabo, R. Can Scenario Planning Catalyse Transformational Change? Evaluating a Climate Change Policy Case Study in Mali. *Futures* **2018**, *96*, 44–56. [CrossRef]
38. Wiebe, K.; Zurek, M.; Lord, S.; Brzezina, N.; Westhoek, H. Scenario Development and Foresight Analysis: Exploring Options to Inform Choices. *Annu. Rev. Environ. Resour.* **2018**, *43*, 545–570. [CrossRef]
39. Amer, M.; Daim, T.U.; Jetter, A. A review of scenario planning. *Futures* **2013**, *46*, 23–40. [CrossRef]
40. Rounsevell, M.D.; Metzger, M.J. Developing qualitative scenario storylines for environmental change assessment. *Wiley Interdiscipl. Rev. Climate Chang.* **2010**, *1*, 606–619. [CrossRef]
41. Johnson, K.A.; Dana, G.; Reich, P.B. Using participatory scenarios to stimulate social learning for collaborative sustainable development. *Ecol. Soc.* **2012**, *17*, 9. [CrossRef]
42. Mistry, J.; Tschirhart, C.; Bovolo, I. Our Common Future? Cross-Scalar Scenario Analysis for Social–Ecological Sustainability of the Guiana Shield, South America. *Environ. Sci. Policy* **2014**, *44*, 126–148. [CrossRef]
43. Biggs, R.; Bohensky, E.; Fabricius, C.; Lynam, T.; Misselhorn, A.; Musvoto, C.; Mutale, M.; Reyers, B.; Scholes, R.J.; Shikongo, S. *Nature Supporting People: The Southern African Millennium Ecosystem Assessment*; Council for Scientific and Industrial Research (CSIR): Pretoria, South Africa, 2004.
44. Biggs, R.; Raudsepp-Hearne, C.; Atkinson-Palombo, C.; Bohensky, E.; Boyd, E.; Cundill, G.; Fox, H.; Ingram, S.; Kok, K.; Spehar, S.; et al. Linking Futures across Scales: A Dialog on Multiscale Scenarios. *Ecol. Soc.* **2007**, *12*, 17. [CrossRef]
45. Chaudhury, M.; Vervoort, J.; Ainslie, A. Participatory Scenarios as a Tool to Link Science and Policy on Food Security under Climate Change in East Africa. *Reg. Environ. Chang.* **2013**, *13*, 389–398. [CrossRef]
46. Enfors, E.; Gordon, L.; Bossio, D. Making Investments in Dryland Development Work: Participatory Scenario Planning in the Makanya Catchment, Tanzania. *Ecol. Soc.* **2008**, *13*, 42. [CrossRef]
47. Kassa, H.; Campbell, B.; Sandewall, M.; Kebede, M.; Tesfaye, Y.; Dessie, G.; Seifu, A.; Tadesse, M.; Garedew, E.; Sandewall, K.; et al. Building Future Scenarios and Uncovering Persisting Challenges of Participatory Forest Management in Chilimo Forest, Central Ethiopia. *J. Environ. Manag.* **2009**, *90*, 1004–1013. [CrossRef]
48. Ojoyi, M.; Mutanga, O.; Abdel-Rahman, E.M. Scenario-based approach in dealing with climate change impacts in Central Tanzania. *Futures* **2017**, *85*, 30–41. [CrossRef]
49. Reinhardt, J.; Liersch, S.; Abdeladhim, M.A.; Diallo, M.; Dickens, C.; Fournet, S.; Hattermann, F.F.; Kabaseke, C.; Muhumuza, M.; Mul, M.L.; et al. Systematic Evaluation of Scenario Assessments Supporting Sustainable Integrated Natural Resources Management: Evidence from Four Case Studies in Africa. *Ecol. Soc.* **2018**, *23*, 5. [CrossRef]
50. Agarwal, C.; Green, G.M.; Grove, J.M.; Evans, T.P.; Schweik, C.M. *A Review and Assessment of Land-Use Change Models: Dynamics of Space, Time, and Human Choice*; US Department of Agriculture: Washington, DC, USA; Forest Service: Washington, DC, USA, 2002; p. 61.
51. Aguiar, A.P.D.; Collste, D.; Harmáčková, Z.V.; Pereira, L.; Selomae, O.; Galafassi, D.; Van Vuuren, D.; Van Der Leeuw, S. Co-Designing Global Target-Seeking Scenarios: A Cross-Scale Participatory Process for Capturing Multiple Perspectives on Pathways to Sustainability. *Glob. Environ. Chang.* **2020**, *65*, 102198. [CrossRef]
52. Kebede, A.S.; Nicholls, R.J.; Allan, A.; Arto, I.; Cazcarro, I.; Fernandes, J.A.; Hill, C.T.; Hutton, C.W.; Kay, S.; Lázár, A.N.; et al. Applying the Global RCP-SSP-SPA Scenario Framework at Sub-National Scale: A Multi-Scale and Participatory Scenario Approach. *Sci. Total Environ.* **2018**, *635*, 659–672. [CrossRef]
53. Rosa, I.M.; Pereira, H.M.; Ferrier, S.; Alkemade, R.; Acosta, L.A.; Akcakaya, H.R.; Den Belder, E.; Fazel, A.M.; Fujimori, S.; Harfoot, M.; et al. Multiscale scenarios for nature futures. *Nat. Ecol. Evolut.* **2017**, *1*, 1416–1419. [CrossRef] [PubMed]
54. Verburg, P.H.; Erb, K.H.; Mertz, O.; Espindola, G. Land System Science: Between Global Challenges and Local Realities. *Curr. Opin. Environ. Sustain.* **2013**, *5*, 433–437. [CrossRef]
55. Palazzo, A.; Vervoort, J.M.; Zougmore, R. Linking Regional Stakeholder Scenarios and Shared Socioeconomic Pathways: Quantified West African Food and Climate Futures in a Global Context. *Glob. Environ. Chang.* **2017**, *45*, 227–242. [CrossRef]
56. Ravera, F.; Tarrasón, D.; Simelton, E. Envisioning adaptive strategies to change: Participatory scenarios for agropastoral semiarid systems in Nicaragua. *Ecol. Soc.* **2011**, *16*, 20. [CrossRef]
57. Ross, D.C. *Mozambique Rising. Building a New Tomorrow*; African Department, International Monetary Fund: Washington, DC, USA, 2014.
58. World Bank. *Mozambique Economic Update, October 2018: Shifting to More Inclusive Growth*; World Bank: Washington, DC, USA, 2018.
59. MICOA (Ministério para a Coordenação da Acção Ambiental). *Relatório do Estudo de Avaliação da Interacção entre a Biodiversidade e Pobreza em Moçambique. Final Report*; MICOA: Maputo, Mozamb, 2008; 139p.
60. Governo do Moçambique. *Estratégia Nacional de Desenvolvimento (2015–2035)*; República De Moçambique: Maputo, Mozambique, 2014; 60p.
61. *World Bank Data. Worldbank.Org/Country/Mozambique—Data: 2008–2018*; World Bank: Washington, DC, USA, 2019.

62. Governo do Moçambique (GdM). Lei Do Orçamento Do Estado Para 2019. In *Ministério da Economia e Finança*; Governo do Moçambique (GdM): Maputo, Mozambique, 2018.
63. World Travel & Tourism Council (WTTC). *Travel & Tourism Economic Impact 2018. Mozambique*; WTTC: London, UK, 2018; 24p. Available online: https://www.wttc.org/economic-impact/country-analysis/country-reports/ (accessed on 1 September 2021).
64. MITADER (Ministério da Terra, Ambiente e Desenvolvimento Rural). *Inventário Florestal Nacional*; Direcção Nacional de Florestas: Maputo, Mozambique, 2018.
65. Governo do Moçambique (GdM). Anuário de Estatísticas Agrárias 2015. In *Direcção de Planificação e Cooperação Internacional. Ministério da Agricultura e Segurança Alimentar*; Governo do Moçambique (GdM): Maputo, Mozambique, 2015.
66. Governo do Moçambique (GdM). Plano estrategico para o desenvolvimento do sector agrario (PEDSA) 2011–2020. In *Ministerio da Agricultura*; Governo do Moçambique: Maputo, Mozambique, 2011.
67. Instituto Nacional de Gestão de Calamidades (INGC). *Main Report: INGC Climate Change Report: Study on the Impact of Climate Change on Disaster Risk in Mozambique*; Asante, K., Brundrit, G., Epstein, P., Fernandes, A., Marques, M.R., Mavume, A., Metzger, M., Patt, A., Queface, A., Sanchez del Valle, R., et al., Eds.; INGC: Maputo, Mozambique, 2009; 338p.
68. Governo do Moçambique (GdM). *Resultados Preliminares de Recolha Estatística, Referentes ao IV Recenseamento Geral da População e Habitação, CENSO 2017*; Governo do Moçambique (GdM): Maputo, Mozambique, 2018.
69. Instituto Nacional de Estatistica (INE). *Projeccoes 2017–2050. Census Data 2017*; Governo do Moçambique: Maputo, Mozambique, 2018.
70. Luz, A.C.; Baumert, S.; Fisher, J.; Grundy, I.; Matediane, M.; Patenaude, G.; Ribeiro, N.; Ryan, C.; Vollmer, F.; Woollen, E.; et al. Charcoal Production and Trade in Southern Mozambique: Historical Trends and Present Scenarios. In Proceedings of the XIV World Forestry Congress, Durban, South Africa, 7–11 September 2015.
71. Devitt, C.; Tol, R.S. Civil War, Climate Change, and Development: A Scenario Study for Sub-Saharan Africa. *J. Peace Res.* **2012**, *49*, 129–145. [CrossRef]
72. Geneletti, D. Environmental Assessment of Spatial Plan Policies through Land Use Scenarios: A Study in a Fast-Developing Town in Rural Mozambique. *Environ. Impact Assess. Rev.* **2012**, *32*, 1–10. [CrossRef]
73. Metzger, M. *Climate Change Historical and Baseline Analysis*; Asante, K., Brundrit, G., Epstein, P., Fernandes, A., Brito, R., Eds.; Study on the Impact of Climate Change on Disaster Risk in Mozambique, INGC Synthesis Report on Climate Change—First Draft; National Institute for Disaster Management: Maputo, Mozambique, 2009.
74. Reed, M.S. Who's in and Why? A Typology of Stakeholder Analysis Methods for Natural Resource Management. *J. Environ. Manag.* **2009**, *90*, 1933–1949. [CrossRef] [PubMed]
75. Hagemann, N.; Zanden, E.H.; Willaarts, B.A.; Holzkämper, A.; Volk, M.; Schönhart, M. Bringing the Sharing-Sparing Debate down to the Ground—Lessons Learnt for Participatory Scenario Development. *Land Use Policy* **2020**, *91*, 104262. [CrossRef]
76. Ritchey, T. Modelling Alternative Futures with General Morphological Analysis. *World Future Rev.* **2011**, *3*, 83–94. [CrossRef]
77. Zwicky, F. The Morphological Approach to Discovery, Invention, Research and Construction. In *New Methods of Thought and Procedure*; Springer: Berlin/Heidelberg, Germany, 1967; pp. 273–297.
78. Slaughter, R.A. Futures for the Third Millennium–Enabling the Forward View. *Syd. Prospect* **2000**, *2*, 369–370.
79. Zorrilla-Miras, P.; Matediane, J.; Mahamane, M.; Nhantumbo, I.; Varela, R.; Metzger, M.J.; Patenaude, G. *Scenarios of Future Land Use Change in Mozambique (2014 and 2015)*; NERC Environmental Information Data Centre: Lancaster, UK, 2018. [CrossRef]
80. F.A.O.; Gumbo, D.J.; Dumas-Johansen, M.; Muir, G.; Boerstler, F.; Xia, Z. *Sustainable Management of Miombo Woodlands—Food Security, Nutrition and Wood Energy*; Food and Agriculture Organization of the United Nations: Rome, Italy, 2018.
81. Iwaniec, D.M.; Cook, E.M.; Davidson, M.J. The Co-Production of Sustainable Future Scenarios. *Landsc. Urban Plan.* **2020**, *197*, 103744. [CrossRef]
82. Jiren, T.; Hanspach, J.; Schultner, J.; Fischer, J.; Bergsten, A.; Senbeta, F.; Hylander, K.; Dorresteijn, I. Reconciling Food Security and Biodiversity Conservation: Participatory Scenario Planning in Southwestern Ethiopia. *Ecol. Soc.* **2020**, *25*, 24. [CrossRef]
83. Baumert, S.; Fisher, J.; Ryan, C.; Woollen, E.; Vollmer, F.; Artur, L.; Mahamane, M. Forgone Opportunities of Large-Scale Agricultural Investment: A Comparison of Three Models of Soya Production in Central Mozambique. *World Dev. Perspect.* **2019**, *16*, 100145. [CrossRef]
84. Hall, R.; Edelman, M.; Borras, S.M.; Scoones, I.; White, B.; Wolford, W. Resistance, Acquiescence or Incorporation? An Introduction to Land Grabbing and Political Reactions from Below. *J. Peasant Stud.* **2015**, *42*, 467–488. [CrossRef]
85. Ndi, F.A. Land Grabbing, Gender and Access to Land: Implications for Local Food Production and Rural Livelihoods in Nguti Sub-Division, South West Cameroon. *Can. J. Afr. Stud./Rev. Canadienne des Études Africaines* **2019**, *53*, 131–154. [CrossRef]
86. Lord, S.; Helfgott, A.; Vervoort, J.M. Choosing diverse sets of plausible scenarios in multidimensional exploratory futures techniques. *Futures* **2016**, *77*, 11–27. [CrossRef]
87. Zurek, M.B.; Henrichs, T. Linking scenarios across geographical scales in international environmental assessments. *Technol. Forecast. Soc. Chang.* **2007**, *74*, 1282–1295. [CrossRef]
88. Lebel, L.; Thongbai, P.; Tianxiang, Y. Subglobal Scenarios. In *Ecosystems and Human Well-Being. Volume 4: Multiscale assessments. Findings of the Subglobal Assessments Working Group of the Millennium Ecosystem Assessment*; Capistrano, D., Samper, C.K., Raudsepp-Hearne, C., Eds.; Island Press: Washington, DC, USA, 2005; pp. 227–258.
89. Allan, A.; Barbour, E.; Nicholls, R.J.; Hutton, C.; Lim, M.; Salehin, M.; Rahman, M.M. Developing Socio-Ecological Scenarios: A Participatory Process for Engaging Stakeholders. *Sci. Total Environ.* **2022**, *807*, 150512. [CrossRef] [PubMed]

90. Smart, T.; Hanlon, J. *Chickens and Beer: Recipe for Agricultural Growth in Mozambique*; Open University/Ciedima: Maputo, Mozambique, 2014.
91. Vollmer, F.; Zorrilla-Miras, P.; Patenaude, G. Charcoal Income as a Means to a Valuable End: Scope and Limitations of Income from Rural Charcoal Production to Alleviate Acute Multidimensional Poverty in Mabalane District, Southern Mozambique. *World Dev. Perspect.* **2017**, *7*, 43–60. [CrossRef]
92. Norfolk, S.; Quan, J.; Mullins, D. *Options for Securing Tenure and Documenting Land Rights in Mozambique: A Land Policy & Practice Paper*; A LEGEND Publication; Natural Resources Institute, University of Greenwich: Chatham, UK, 2020.
93. Chardonnet, B. *Africa Is Changing: Should Its Protected Areas Evolve? Reconfiguring the Protected Areas in Africa*; FRANCE-IUCN PARTNERSHIP, International Union for the Conservation of Nature and Natural Resources: Gland, Switzerland, 2019; 32p.
94. Ferraro, P.J.; Hanauer, M.M.; Sims, K.R. Conditions Associated with Protected Area Success in Conservation and Poverty Reduction. *Proc. Natl. Acad. Sci. USA* **2011**, *108*, 13913–13918. [CrossRef] [PubMed]
95. Bruner, A.G.; Gullison, R.E.; Balmford, A. Financial Costs and Shortfalls of Managing and Expanding Protected-Area Systems in Developing Countries. *BioScience* **2004**, *54*, 1119–1126. [CrossRef]
96. Mahamane, M.; Zorrilla-Miras, P.; Baumert, S. Understanding Land Use, Land Cover and Woodland-Based Ecosystem Services Change, Mabalane, Mozambique. *Energy Environ. Res.* **2017**, *7*, 1–22. [CrossRef]
97. Governo de Moçambique. *Política Florestal e Estratégia da sua Implementação 2020–2035*; Ministério da Terra e Ambiente, República de Moçambique, Governo de Moçambique: Maputo, Mozambique, 2020.
98. Governo do Moçambique. *Intended Nationally Determined Contribution of Mozambique to the United Nations Framework Convention on Climate Change (UNFCCC)*; Governo de Moçambique (GdM): Maputo, Mozambique, 2018.
99. Silva, J.A.; Sedano, F.; Flanagan, S.; Ombe, Z.A.; Machoco, R.; Meque, C.H.; Sitoe, A.; Ribeiro, N.; Anderson, K.; Baule, S.; et al. Charcoal-Related Forest Degradation Dynamics in Dry African Woodlands: Evidence from Mozambique. *Appl. Geogr.* **2019**, *107*, 72–81. [CrossRef]
100. Sedano, F.; Lisboa, S.N.; Tucker, C.J. Monitoring Forest Degradation from Charcoal Production with Historical Landsat Imagery. A Case Study in Southern Mozambique. *Environ. Res. Lett.* **2019**, *15*, 015001. [CrossRef]
101. Bey, A.; Jetimane, J.; Lisboa, S.N.; Ribeiro, N.; Sitoe, A.; Meyfroidt, P. Mapping Smallholder and Large-Scale Cropland Dynamics with a Flexible Classification System and Pixel-Based Composites in an Emerging Frontier of Mozambique. *Remote Sens. Environ.* **2020**, *239*, 111611. [CrossRef]
102. Bleyer, M.; Kniivilä, M.; Horne, P.; Sitoe, A.; Falcão, M.P. Socio-economic impacts of private land use investment on rural communities: Industrial forest plantations in Niassa, Mozambique. *Land Use Policy* **2016**, *51*, 281–289. [CrossRef]
103. Newitt, M.D.D. *A History of Mozambique*; Indiana University Press: Bloomington, IN, USA, 1995.
104. Temudo, M.P.; Silva, J.M. Agriculture and forest cover changes in post-war Mozambique. *J. Land Use Sci.* **2012**, *7*, 425–442. [CrossRef]
105. Unruh, J. Land tenure and identity change in postwar Mozambique. *Geojournal* **1998**, *46*, 89–99. [CrossRef]
106. Vermeulen, S.; Cotula, L. Over the heads of local people: Consultation, consent, and recompense in large-scale land deals for biofuels projects in Africa. *J. Peasant Stud.* **2010**, *37*, 899–916. [CrossRef]
107. Zaehringer, J.G.; Atumane, A.; Berger, S.; Eckert, S. Large-scale agricultural investments trigger direct and indirect land use change: New evidence from the Nacala corridor, Mozambique. *J. Land Use Sci.* **2018**, *13*, 325–343. [CrossRef]
108. UNESCO Institute for Statistics (Data Source: uis.unesco.org). Available online: https://data.worldbank.org/indicator/SE.PRM.ENRL.TC.ZS?locations=MZ (accessed on 8 August 2018).
109. Available online: http://data.worldbank.org/country/mozambique (accessed on 8 August 2018).
110. *Plano Operacional para o Desenvolvimento Agrário de Moçambique (2015–2019)*; Ministério da Agricultura e Segurança Alimentar, República de Moçambique: Maputo, Mozambique, 2015.
111. Available online: https://data.worldbank.org/indicator/IT.CEL.SETS.P2?locations=MZ&year_high_desc= (accessed on 8 August 2018).
112. Marzoli, A. *Inventário Florestal Nacional: Avaliação Integrada da Floresta em Moçambique (AIFM)*; Direção Nacional de Terras e Florestas: Maputo, Mozambique, 2007.
113. Direção Nacional de Terras e Florestas. *Inventário Florestal Nacional*; Ministério da Terra, Ambiente e Desenvolvimento Rural, República De Moçambique: Maputo, Mozambique, 2018.
114. Ryan, C.M.; Berry, N.J.; Joshi, N. Quantifying the causes of deforestation and degradation and creating transparent REDD+ baselines: A method and case study from central Mozambique. *Appl. Geogr.* **2014**, *53*, 45–54. [CrossRef]
115. Available online: https://data.worldbank.org/country/mozambique (accessed on 8 August 2018).

Communication

The Future of Rurality: Place Attachment among Young Inhabitants of Two Rural Communities of Mediterranean Central Chile

Paulina Rodríguez-Díaz [1,2], Rocío Almuna [1,3,4], Carla Marchant [5], Sally Heinz [1], Roxana Lebuy [1], Juan L. Celis-Diez [6,7] and Pablo Díaz-Siefer [1,6,*]

1 Centro Regional de Investigación e Innovación para la Sostenibilidad de la Agricultura y los Territorios Rurales_CERES, Quillota 2260000, Chile; parodriguez6@uc.cl (P.R.-D.); rocio.am92@gmail.com (R.A.); sallyheinza@gmail.com (S.H.); rlebuy@centroceres.cl (R.L.)
2 Programa de Doctorado en Geografía, Instituto de Geografía, Pontificia Universidad Católica de Chile, Santiago 8320000, Chile
3 ECOS (Ecosystem-Complexity-Society) Co-Laboratory, Center for Local Development (CEDEL) & Center for Intercultural and Indigenous Research (CIIR), Villarrica Campus, Pontificia Universidad Católica de Chile, Villarrica 4930000, Chile
4 School of Agriculture and Environment, Albany Campus, The University of Western Australia, Albany 6330, Australia
5 Instituto de Ciencias Ambientales y Evolutivas, Laboratorio de Estudios Territoriales LAbT UACh, Universidad Austral de Chile, Valdivia 5090000, Chile; carla.marchant@uach.cl
6 Escuela de Agronomía, Pontificia Universidad Católica de Valparaíso, Valparaíso 2340000, Chile; juan.celis@pucv.cl
7 Centro de Acción Climática, Pontificia Universidad Católica de Valparaíso, Santiago 8320000, Chile
* Correspondence: pablo.diaz@pucv.cl

Citation: Rodríguez-Díaz, P.; Almuna, R.; Marchant, C.; Heinz, S.; Lebuy, R.; Celis-Diez, J.L.; Díaz-Siefer, P. The Future of Rurality: Place Attachment among Young Inhabitants of Two Rural Communities of Mediterranean Central Chile. *Sustainability* **2022**, *14*, 546. https://doi.org/10.3390/su14010546

Academic Editor: John McDonagh

Received: 18 September 2021
Accepted: 14 December 2021
Published: 5 January 2022

Publisher's Note: MDPI stays neutral with regard to jurisdictional claims in published maps and institutional affiliations.

Abstract: Rural livelihoods are under threat, not only from climate change and soil erosion but also because young people in rural areas are increasingly moving to urbanized areas, seeking employment and education opportunities. In the Valparaiso region of Chile, megadrought, soil degradation, and industrialization are driving young people to leave agricultural and livestock activities. In this study, our main objective was to identify the factors influencing young people living in two rural agricultural communities (Valle Hermoso and La Vega). We conducted 90 online surveys of young people aged 13–24 to evaluate their interest in living in the countryside (ILC). We assessed the effect of community satisfaction, connectedness to nature, and social valuation of rural livelihoods on the ILC. The results show that young people were more likely to stay living in the countryside when they felt satisfied and safe in their community, felt a connection with nature, and were surrounded by people who enjoyed the countryside. These results highlight the relevance of promoting place attachment and the feeling of belonging within the rural community. Chilean rural management and local policies need to focus on rural youth and highlight the opportunities that the countryside provides for them.

Keywords: rural exodus; rural livelihood; rural migration; rural youth; belonging

1. Introduction

The depopulation of rural areas is a demographic phenomenon worldwide [1–6] and is particularly relevant in Latin American countries [7], with societies shifting from agrarian to urban-industrial economies [8]. Previous studies show that the main factors driving the rural exodus are employment, recreational and education opportunities, together with the local effects of climate change and the decrease in agricultural productivity, due to the degradation and loss of soil fertility [1,9,10]. In this context, rural youth is a more geographically mobile demographic than rural adults [11]. Thus, the increasing exodus of young people from rural to urban areas threatens the stability of rural livelihoods [1,12,13].

Despite the aforementioned threats to agricultural productivity, rural areas are generally considered an attractive place to live due to their social and natural environment, and high quality of life [14–16]. In fact, rural youth who have a strong attachment to a place due to close relationships between community members, family and friends are most inclined to remain in their rural locality [10,17,18]. Altman and Low [19] assigned the term "place attachment" to the action of developing an emotional bond with places, consequently generating a bond with the physical and socio-cultural environment in which inhabitants develop their daily activities and personal experiences—this is considered a feel-good factor [20]. People develop these attachments to places where they feel secure and protected, and that they consider to be their home [21]. According to Riethmuller et al. [22], together with economic, educational, and interpersonal factors, place attachment is relevant to consider when addressing the motives for migration among rural young people, as well as their interest in returning to their rural roots.

In Latin America, the world's most urbanized region [7,23,24], rural youth face significant disadvantages and poverty levels higher than those of their urban counterparts, placing them as one of the most vulnerable social groups [25]. This scenario favors the exodus of the young rural population to urban areas [26,27]. In this context, Chile is no exception, since the centralism of the political–administrative system, along with the neoliberal economic model, has mainly promoted the tertiary and financial sectors. This has led to a lack of opportunities for young people in rural settlements and has had a detrimental effect on primary activities in rural areas, contributing to the urban–rural territorial imbalance [28]. An example of this is the centralized education system that drives young people to leave their rural homes and migrate to cities, which, in the long term, has a strong influence on definitive migration to urban centers [29].

In the Valparaiso region of central Chile, migration patterns can be influenced by factors such as water scarcity and the demographic aging process, among others [30]. According to the National Institute of Statistics [31], demographic aging, when projected to 2035, will be high in the Valparaiso region, with 22.2% of the population being over 65 years of age, making it the second-oldest demographic region in the country. Furthermore, from 2002 to 2017, the intercensal variation of the rural population between 13 and 24 years old in the Valparaiso region was −4.1% [32]. Therefore, the sustainability of peasant family farming is under threat.

The motivation for this study emerged from workshops on a soil restoration project for degraded slopes, which was implemented with the assistance of two rural agricultural communities in the Valparaiso region (Valle Hermoso and La Vega), as part of the outreach activities of a research and development project. In these workshops, the communities voiced concerns about youth migration. Although young people seem to have a propensity for migration, as has been the trend in Latin America since the mid-twentieth century, different studies have underlined the desire shared by rural adults and youth to maintain rural continuity [33–36]. In the cases of the Valle Hermoso and La Vega communities, the conformation of the peasantry, as well as the development of local ancestral knowledge, are the result of a process of in situ cultural syncretism between creole and indigenous traditions. Additionally, the imminent disappearance of rural livelihoods is a relevant matter of concern among older peasants. Because of their historical dependence on agriculture, their cultural and emotional relationship with land ownership, and the rapid changes they are currently experiencing, these rural communities present an interesting case study to explore rural youth's interest in staying in the countryside, especially when considering that, on the one hand, Chile is one of the countries with the highest urbanization rates within the already urbanized Latin America [7]. On the other hand, Chile has an economic model based on international trade, which confers an excellent study model by which to project migration patterns in an increasingly globalized world. Therefore, the aim of this study is to identify the main factors that influence young rural people from Valle Hermoso and La Vega in their choice to live in the countryside. We target one specific question: what factors of place attachment should be encouraged to avoid youth depopulation?

2. Literature Review

Understanding the characteristics of the youth that live in rural areas is essential to the future of these areas. Pardo [37] points out that the main characteristic of rural young people today is a higher educational level that makes them more flexible and open to innovation, better able to use new technologies, forms of socialization, and methods of acquisition of knowledge, making them relevant actors in their territories. However, their vision, voices, and interests have not been solicited, either in public policies or in the construction of dynamics for development. According to Díaz and Fernández [38], in Latin American countries, rural youth are in a situation of greater vulnerability concerning their urban peers, having fewer job opportunities for non-precarious employment and fewer possibilities of access to education, as well as higher rates of poverty. This generates a higher proportion of people who neither work nor study, especially in the case of young rural women.

2.1. Community, Rural Livelihoods and Nature as Motivations for Living in the Countryside

Previous studies have shown that the interest of rural youth in living in the countryside is related to both economic [39] and non-economic factors [40], in addition to structural and cultural factors [9]. Thus, the motivations of young people to migrate from or to return to rural places after completing their studies are influenced by family pressure, employment expectations, quality of life, personal background, environmental impacts on agriculture, lack of resources, and the local community environment [41,42].

People in rural areas tend to develop stronger attachments to their community than those in urban areas; therefore, leaving their community for educational or employment opportunities can be very difficult [22,43]. In this sense, community satisfaction is one of the factors influencing the young population exodus that, paradoxically, is affected by the same out-migration. The out-migration of young people can reduce the opportunities for social interaction for those who stay. Young people have less involvement in community organizations, sports and church groups, and other initiatives that nourish the local social capital [44,45]. This decline in the young population may harm the community satisfaction of the young people who stay.

The rural world has privileges that can impact young people's decision to develop their life projects in the countryside [46,47]. This has mainly been influenced by connectivity improvements and technical skills that facilitate mobility [48] and the rise of remote working triggered by the COVID-19 crisis. The pandemic has promoted an increased interest in rural spaces close to nature [49]. However, high rates of poverty and difficulties in access to certain goods and services in the countryside often make this alternative unattractive for many. For example, rural youth in economically distressed places must develop their plans in a context where educational and employment opportunities are generally found elsewhere, creating the need to migrate [50].

Furthermore, connectedness to nature, when understood as a personal attitude [51], is considered one of the multiple and dynamic dimensions of place attachment. Generally, young people express a positive attitude toward nature as an attractive aspect of rural places and positively link such places to their geographical background [52]. Haukanes [53] found that the arguments from those who prefer to live in the countryside are primarily based on the concept of the rural idyll, where the beauty of nature, tranquility, and air quality play a central role. However, in the scientific literature, the connectedness to nature is not usually considered an influential factor in the migration decision of rural youth. Our study aims to quantitatively assess whether rural youth's social and natural environment are factors that influence their interest in living in the countryside.

2.2. Rural Development National Policy

To date, in Chile, there is almost no territorial planning in rural areas [28], which makes it difficult to implement public policies that contribute to the development of the countryside and promote the interest of young people in staying. However, in 2018, new legislation on territorial planning was approved under law No. 21074 of the Government of Chile on strengthening regionalization. This laid the ground for a new approach to land-use planning, which involves the preparation of the National Rural Development Policy [28]. By this law, all instruments for territorial planning must be subject to this policy's requirements.

The National Rural Development Policy was approved and published in 2020 [54]. Its goal is to "*improve the quality of life and increase the opportunities of the population living in rural territories, generating appropriate conditions for their integral development, through the gradual, planned and sustained adoption of a paradigm that conceives a public action with a territorial approach and that fosters synergies between public, private and civil society initiatives. In this way, the National Rural Development Policy expects to contribute to a greater territorial balance in the country, promoting the sustainable development of its smaller populated settlements*". It focuses on four main areas: social goods; economic opportunities; environmental sustainability; and culture and identity. It is focused on the entire national rural territory, so all the territorial planning instruments must be in line with it. It includes a definition of rural territory, objectives and guiding principles, and the identification of key areas and strategic lines for rural development. It should be noted that this policy contemplates the strengthening, development, and articulation of programs and instruments that seek to satisfy the needs of groups that require priority attention, including children and youth.

3. Materials and Methods

3.1. Study Site

The study was performed in two agricultural communities in the Mediterranean zone of central Chile, specifically in the Valparaiso region: Valle Hermoso, located in the La Ligua district, and La Vega located in the Olmué district (see Figure 1). Both communities have similar characteristics, such as being agricultural communities whose origins date back to pre-Hispanic times [55,56]. Likewise, both places have shown a gradual abandonment of agriculture and livestock as the main economic activities. Another coincidence to note is the presence of a demographic dynamic characterized by a decrease in the young population: from 2002 to 2017, the intercensal variation of the rural population between 13 and 24 years old was −1.6% in La Ligua and −0.5% in Olmué [33]. Although the intercensal variation of the districts does not represent the region's trend, it should be noted that today, there are only 7 agricultural communities in the Valparaíso region that are undergoing a process of territorial fragmentation [57].

3.2. Sampling

We conducted an online survey (using Google Forms) among young people from 13 to 24 years old, to understand their interest in living and remaining in the countryside. The relevant fieldwork was conducted between February and March 2021. We involved both communities in order to maximize the sample size, since agricultural communities in the Valparaiso region are few in number and small in population. We calculated the sample size using Bartlett et al.'s [58] method for categorical data, considering an alpha level of $\alpha = 0.1$ and t-value = 1.65 as a desired level of precision (as used by, e.g., [59]). This calculation estimated a sample size of 58 rural young people. However, we decided to collect at least 100 responses to utilize a more conservative approach. Community leaders and members provided us with a list of 61 young people, who were contacted directly via phone call, and the survey was sent via WhatsApp. Furthermore, we also used a chain-sampling method among the members of the communities in order that the survey could reach more young people. It should be noted that we started with physical surveys, but due to the COVID-19 pandemic, we adapted to an online format.

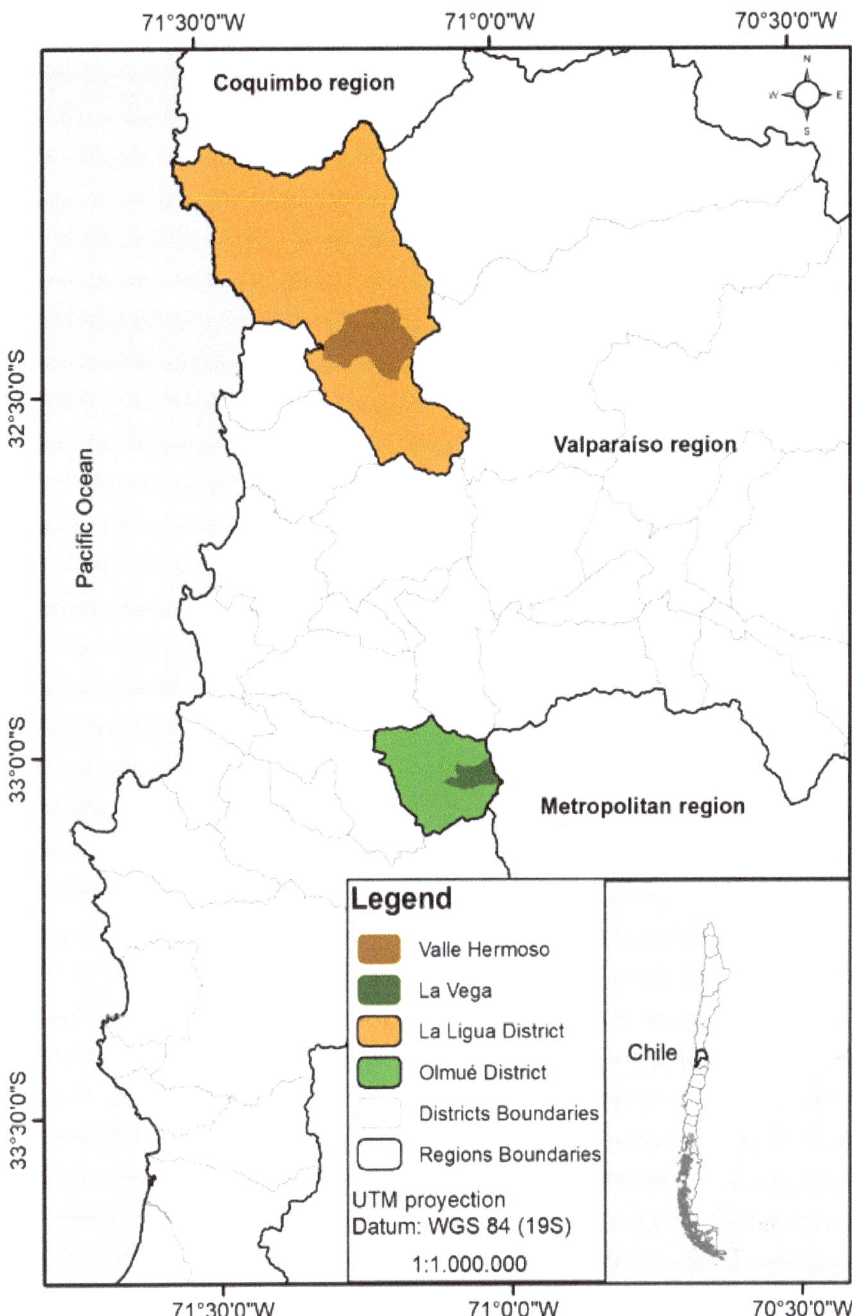

Figure 1. Location of the two rural communities from Mediterranean central Chile.

3.3. Measures

To measure the main factors that influence young people to live in the countryside, we designed a survey with two main sections: (i) demographic questions to gather information on the socioeconomic and cultural profile of the respondents; (ii) questions related to the interest of rural young people in living in the countryside. The survey instrument was piloted on 17 young volunteers from a different rural community to those taking part in this study in order to adjust timing and to assess the respondents' understanding of the individual items.

The first section had 20 data points relating to their personal information (age, gender, education, place of residence, and studies), family characteristics (parents' educational level and family's economic activity), and their participation in farm-related work. In the second section, we designed 4 different 5-point Likert scales with 36 items in total to assess the interest of rural youth in living in the countryside (Supplementary Material Figure S1). The 4 scales were: (1) interest in living in the countryside (hereafter, ILC) related to the respondents' projection to live in the countryside; (2) community satisfaction (hereafter, CS), related to the level of satisfaction of the rural youth with the people, rural environment, and rural livelihoods of their community; (3) connectedness to nature (hereafter, CN), associated with the affective and physical relationship that rural youth have with the natural world; (4) social valuation of rural livelihoods (hereafter, SVRL), related to the influence exerted by the social environment over young rural people (see Supplementary Material Table S1 for details of the items of each scale). The wording of some items was modified to facilitate the respondents' understanding and to better suit Chilean idiosyncrasies.

3.4. Data Analysis

All the statistical analyses were conducted using the statistical software R v4.0.4 [60], considering a significance level of 0.05. First, we used Cronbach's alpha reliability analysis to estimate the internal consistency of each scale in Section 2, which were ILC, CS, CN, and SVRL, using the *psych* package [61]. We treated the scale scores as ordinal data; therefore, we performed nonparametric statistics to analyze these scores [62,63]. Thus, to establish which factors of place attachment should be encouraged to avoid youth depopulation, Spearman's rho correlation coefficients were used to analyze the association between the ILC and the other variables of interest. Finally, a nonparametric quantile regression (hereafter, NQR) was performed with ILC as a dependent variable and CS, CN, and SVRL as predictors, using the *quantreg* package [64]. We chose to use NQR because it allows for the estimation at various quantiles of the dependent variable, rather than presuming a uniform mean effect, without making assumptions about the distribution of the dependent variable [65].

4. Results

4.1. Socioeconomic and Cultural Characteristics of Respondents

We received 106 responses; however, 16 were excluded due to erroneous duplication ($n = 11$) and ages out of range ($n = 5$), leaving a final sample of 90 young people from the two communities. It should be noted that we produced three physical surveys. The average age of the respondents was 18 ± 3.2 years, with 50% women ($n = 45$) and 50% men ($n = 45$). In addition, the respondents tended to develop their studies within the place where they grew up (40%), or in another place that was still within the Valparaiso region (38.9%). The majority of respondents had never lived outside their rural community (70%); those who had lived beyond it had done so mainly due to their studies. The families were mostly engaged in agriculture (51.1%) and textile production (36.7%). Regarding countryside activities, 84.4% fed and cared for animals, 64.4% harvested, and 57.7% prepared the soil and sowed (Supplemental Material Table S2 and Figure S2).

4.2. Interest in Living in the Countryside

All measure scales showed good reliability: ILC (Cronbach's $\alpha = 0.77$), CS (Cronbach's $\alpha = 0.71$), CN (Cronbach's $\alpha = 0.71$) and SVRL (Cronbach's $\alpha = 0.65$). Respondents showed a mean score ± SE of 3.66 ± 0.01 for ILC, 3.67 ± 0.06 for CS, 4.46 ± 0.06 for CN and 3.72 ± 0.05 for SVRL.

Spearman's correlation coefficients, between ILC and the other three scales, are presented in Table 1. All correlations showed moderate strength. The results show that higher scores on ILC were positively related to higher scores on CS, and the same occurred between CN and SVRL (all were $p < 0.01$). In addition, CS, CN, and SVRL are related to each other.

Table 1. Spearman correlation tests between ILC and CS, CN and SVRL.

Variable	ILC	CS	CN	SVRL
ILC				
CS	0.404 **			
CN	0.524 **	0.432 **		
SVRL	0.480 **	0.447 **	0.336 **	

Abbreviations: ILC: interest in living in the countryside; CS: community satisfaction; CN: connectedness to nature; SVRL: social valuation of rural livelihoods. ** Represents statistical significance at the 0.01 level.

The results of the NQR between ILC and each factor are provided in Table 2. The NQR showed that there is a positive effect of CS on all the quantiles of ILC; thus, if a young person feels satisfied living within their community, or if they feel safe living in it, they will show more interest in staying in the countryside. Similarly, CN had a positive effect on all quantiles of ILC, reflecting that the more connected a young person is to their natural surroundings and biodiversity, the more interest they will show in living in the countryside. Finally, there is a positive effect of SVRL on all quantiles of ILC, meaning that the social environment has an influence on young people; specifically, when there is a transfer of knowledge about field activities, or when friends find it appealing to stay in the countryside, this contributes to the young people's desire to live near them or practice the same activities. Therefore, the effect of the three factors on interest in living in the countryside is constant across the conditional distribution of ILC. In addition, the effect of the three factors on ILC showed an asymmetric dependence structure with a left-tail dependence (i.e., the factors have a greater influence on the lower quantiles of the ILC).

Table 2. Nonparametric quantile regressions, estimated at the 0.2, 0.4, 0.6, and 0.8 quantiles of ILC. The value in parentheses represents the standard error (SE).

Variable	ILC			
	Q0.2 (SE)	Q0.4 (SE)	Q0.6 (SE)	Q0.8 (SE)
CS	1.25 (0.31) **	1.25 (0.31) **	1.25 (0.31) **	1.25 (0.31) **
CN	0.59 (0.27) *	0.59 (0.27) *	0.59 (0.27) *	0.59 (0.27) *
SVRL	0.63 (0.26) **	0.63 (0.26) **	0.63 (0.26) **	0.63 (0.26) **

Abbreviations: ILC: interest in living in the countryside; CS: community satisfaction; CN: connectedness to nature; SVRL: social valuation of rural livelihoods. * Represents statistical significance at the 0.05 level. ** Represents statistical significance at the 0.01 level.

5. Discussion

Our results show that when young people were satisfied with their community surroundings, they were likely to develop a strong place attachment and, hence, were more interested in living in the countryside [18,66]. In addition, the CS scale is related to people, rural environment, and rural livelihoods within the community. This could explain the positive correlation between CS and CN, and SVRL. Therefore, satisfaction with their social and natural environment could promote a desire in rural youth for this lifestyle. In addition, our study contributes to the few published quantitative studies that consider both the natural and the social environment in order to determine young people's interest in living in the countryside (e.g., [18]).

Social environment strongly influenced place attachment [19], explaining the positive relationship between interest in living in the countryside and social valuation. A rural livelihood comprises the possibilities, assets—including both material and social resources—

and activities necessary to earn a living [46]; thus, the social valuation of rural livelihoods could influence the decision to migrate from the countryside. Similarly, previous studies have shown that the family circle contributes to stimulating the permanence of young people in the countryside [13,41]. Moreover, family and friends are relevant motivations for returning to the local rural communities for those who have migrated [67–70]. Thus, to strengthen our findings, future research could consider rural young people who migrated to cities, because work and family expectations and the identity of young subjects are closely linked to the imaginary ideal of the city [71].

The shared outdoor spaces of a community are essential for the satisfaction of its members, mainly because of the opportunities to visit natural areas and have countryside views [72]. Connectedness to nature was the scale with the highest score (mean $\pm$ SE = 4.46 $\pm$ 0.06). In fact, young people living in rural areas have a strong connection with nature in comparison with young people in urban areas [73–75], probably because rural youth usually spend more time outdoors. Thus, CN was the scale more related to ILC ($r = 0.524$, $p \leqq 0.01$). This strong connection with nature can promote a strong place attachment [52,76,77] and, therefore, a greater interest in living in the countryside. This is in line with previous studies that have shown that nature is an important factor motivating young people to stay in their rural communities, meaning that the natural environment and outdoor recreation have the potential to create a sense of attachment in young people to their homes [14,18,52]. However, today, increased access to screens may have a detrimental effect on the connection with nature among new generations [78,79], which is why future studies should consider the influence of screens and technology on interest in living in the countryside.

The rural landscape, which includes agricultural and natural areas, has a historical–cultural value and, therefore, offers excellent potential for agrotourism, which can become an opportunity for local development [80,81] and help reduce youth migration. Unfortunately, the expansion of the agri-business model, through the extensive replacement of natural areas, not only contributes to the loss of biodiversity but also to the loss of ecosystem services, resulting in a degradation of the landscape and potentially leading to rural youth migration, due to the loss of people's well-being [82]. Thus, access to land often constitutes a barrier for those young people who wish to remain in the countryside and develop an enterprise [83].

Our results can constitute a tool for future rural management and local policies for promoting rural livelihoods, which should focus on the enhancement of the relationship between people, rural livelihoods, and natural areas of the local community in order to increase community satisfaction and, consequently, place attachment. For example, schools could promote a feeling of belonging to the community and the development of a rural identity among young people; municipalities could make rural youth aware of the potentialities of employment and entrepreneurship opportunities in rural areas and promote their local festivals and cultural practices, to create ties within the community. However, this must always be carried out in permanent dialog with the local community. This can certainly mesh with the Chilean Rural Development National Policy [55], which considers children, teenagers, and young adults as being among the priority groups for programs and instruments focused on integrated rural development. In the current context of the global COVID-19 pandemic, this has become particularly urgent. This pandemic has exposed the need for rethinking urban lifestyles and, as rural spaces are increasingly sought after, it is possible that a scenario of greater pressure and disputes for rural territories may emerge in the future.

6. Conclusions

The depopulation of rural areas is an important challenge for the sustainability of rural territories in Latin American countries [1–7]. Our results showed that a young person who feels satisfied in their community, connected to the natural surroundings, and whose family and friends had a positive valuation of the countryside is more interested in the countryside. This contributes to the evidence that both social relationships and connection to the natural environment are important factors in a young person's decision to live in a place similar to where he or she grew up. These results highlight how relevant it is to promote place attachment and the feeling of belonging within the rural community, which translates into young people's interest in living in the countryside. Accordingly, Chilean rural management and local policies should focus not only on economic or employment issues but also on improving young people's social and natural environment valuation to increase their place attachment, promoting local festivals and cultural practices to create ties within the community.

Supplementary Materials: The following are available online at https://www.mdpi.com/article/10.3390/su14010546/s1, Table S1: Scales questionnaire developed in this study, Table S2: Sociodemographic characteristics of respondents ($n = 90$), Figure S1: Main drivers of depopulation of rural areas according to the pull–push approach, Figure S2: Sociodemographic data of our study, locality, and region.

Author Contributions: Conceptualization, P.D.-S.; methodology, P.R.-D., C.M., S.H., R.L. and P.D.-S.; data acquisition, P.R.-D. and S.H.; formal analysis, P.D.-S.; writing—original draft preparation, P.R.-D., R.A. and P.D.-S.; writing—review and editing, P.R.-D., R.A., C.M., S.H., R.L., J.L.C.-D. and P.D.-S. All authors have read and agreed to the published version of the manuscript.

Funding: This research was funded by the Agencia Nacional de Investigación y Desarrollo ANID, Chile (https://www.anid.cl accessed on 5 November 2021) through the grants ANID/REGIONAL/R19F10017 and ANID REGIONAL/CERES/R19A1002. J.L.C.-D. and P.D.-S. were supported by ANID/PIA/ACT192027. P.R.-D. was funded by the ANID doctoral scholarship grant No2121239.

Institutional Review Board Statement: Not applicable.

Informed Consent Statement: Informed consent was obtained from all subjects involved in the study.

Data Availability Statement: Not applicable.

Acknowledgments: The authors acknowledge the collaboration of members of rural communities for their contribution to the survey presented in this article. J.L.C.-D. is an associate researcher at the Institute of Ecology and Biodiversity (IEB).

Conflicts of Interest: The authors declare no conflict of interest. The funders had no role in the design of the study; in the collection, analyses, or interpretation of data; in the writing of the manuscript, or in the decision to publish the results.

References

1. Bjarnason, T.; Thorlindsson, T. Should I stay or should I go? Migration expectations among youth in Icelandic fishing and farming communities. *J. Rural Stud.* **2006**, *22*, 290–300. [CrossRef]
2. Bayona-I-Carrasco, J.; Gil-Alonso, F. Is Foreign Immigration the Solution to Rural Depopulation? The Case of Catalonia (1996–2009). *Sociol. Rural.* **2013**, *53*, 26–51. [CrossRef]
3. Vásquez Wiedeman, C.; Vallejos Quilodrán, D. Migración juvenil rural en la región del Maule, Chile. *Rev. Cienc. Soc.* **2014**, *27*, 91–108.
4. Johnson, K.M.; Lichter, D.T. Rural Depopulation: Growth and Decline Processes over the Past Century. *Rural Sociol.* **2019**, *84*, 3–27. [CrossRef]
5. Rodríguez-Soler, R.; Uribe-Toril, J.; De Pablo Valenciano, J. Worldwide trends in the scientific production on rural depopulation, a bibliometric analysis using bibliometrix R-tool. *Land Use Policy* **2020**, *97*, 104787. [CrossRef]
6. Yeboah, T. Future aspirations of rural-urban young migrants in Accra, Ghana. *Child. Geogr.* **2021**, *19*, 45–58. [CrossRef]
7. Rozzi, R. Biocultural Ethics: From Biocultural Homogenization toward Biocultural Conservation. In *Linking Ecology and Ethics for a Changing World: Ecology and Ethics*; Rozzi, R., Pickett, S., Palmer, C., Armesto, J., Callicott, J., Eds.; Springer: Dordrecht, The Netherlands, 2013; Volume 1, pp. 9–32. [CrossRef]
8. Li, Y.; Westlund, H.; Liu, Y. Why some rural areas decline while some others not: An overview of rural evolution in the world. *J. Rural Stud.* **2019**, *68*, 135–143. [CrossRef]
9. Thissen, F.; Fortuijn, J.D.; Strijker, D.; Haartsen, T. Migration intentions of rural youth in the Westhoek, Flanders, Belgium and the Veenkoloniën, The Netherlands. *J. Rural Stud.* **2010**, *26*, 428–436. [CrossRef]
10. Mclaughlin, D.K.; Shoff, C.M.; Demi, M.A. Influence of Perceptions of Current and Future Community on Residential Aspirations of Rural Youth. *Rural Sociol.* **2014**, *79*, 453–477. [CrossRef]

11. Cazzuffi, C.; Fernández, J. *Rural Youth and Migration in Ecuador, Mexico and Peru*; Serie documento de trabajo N° 235. Programa Jóvenes Rurales, Territorios y Oportunidades: Una estrategia de diálogos de políticas; RIMISP: Santiago, Chile, 2018.
12. Stockdale, A. Out-migration from rural Scotland: The importance of family and social networks. *Sociol. Rural.* **2002**, *42*, 41–64. [CrossRef]
13. Llorent-Bedmar, V.; Cobano-Delgado Palma, V.C.; Navarro-Granados, M. The rural exodus of young people from empty Spain. Socio-educational aspects. *J. Rural Stud.* **2021**, *82*, 303–314. [CrossRef]
14. Gosnell, H.; Abrams, J. Amenity migration: Diverse conceptualizations of drivers, socioeconomic dimensions, and emerging challenges. *GeoJournal* **2011**, *76*, 303–322. [CrossRef]
15. Lichter, D.T.; Brown, D.L. Rural America in an urban society: Changing spatial and social boundaries. *Annu. Rev. Sociol.* **2011**, *37*, 565–592. [CrossRef]
16. Marchant, C.; Rojas, F. Transformaciones locales y nuevas funcionalidades económicas vinculadas a las migraciones por amenidad en la Patagonia chilena. *Rev. Géographie Alp.* **2015**, *105*, 1–19. [CrossRef]
17. Glendinning, A.; Nuttall, M.; Hendry, L.; Kloep, M.; Wood, S. Rural communities and well-being: A good place to grow up? *Sociol. Rev.* **2003**, *51*, 129–156. [CrossRef]
18. Ulrich-Schad, J.D.; Henly, M.; Safford, T.G. The role of community assessments, place, and the great recession in the migration intentions of Rural Americans. *Rural Sociol.* **2013**, *78*, 371–398. [CrossRef]
19. Altman, I.; Low, S. *Place Attachment*; Plenum Press: New York, NY, USA, 1992.
20. Dallimer, M.; Irvine, K.N.; Skinner, A.M.J.; Davies, Z.G.; Rouquette, J.R.; Maltby, L.L.; Warren, P.H.; Armsworth, P.R.; Gaston, K.J. Biodiversity and the feel-good factor: Understanding associations between self-reported human well-being and species richness. *Bioscience* **2012**, *62*, 47–55. [CrossRef]
21. Hernández, B. Place attachment: Antecedents and consequences (Antecedentes y consecuencias del apego al lugar). *Psyecology* **2021**, *12*, 99–122. [CrossRef]
22. Riethmuller, M.L.; Dzidic, P.L.; Newnham, E.A. Going rural: Qualitative perspectives on the role of place attachment in young people's intentions to return to the country. *J. Environ. Psychol.* **2021**, *73*, 101542. [CrossRef]
23. CEPAL. *Población, Territorio y Desarrollo Sostenible*; Naciones Unidas: Santiago, Chile, 2012.
24. Rozzi, R. Biocultural Ethics: From Biocultural Homogenization toward Biocultural Conservation. In *Linking Ecology and Ethics for a Changing World: Values, Philosophy, and Action*; Rozzi, S., Pickett, C., Palmer, J., Armesto, J.J., Callicott, J.B., Eds.; Springer: New York, NY, USA, 2014; pp. 9–32.
25. Guiskin, M. *Situación de las Juventudes Rurales en América Latina y el Caribe*; Serie Estudios y Perspectivas-CEPAL: Mexico City, Mexico, 2019.
26. Sili, M.; Fachelli, S.; Meiller, A. Juventud Rural: Factores que influyen en el desarrollo de la actividad agropecuaria. Reflexiones sobre el caso argentino. *Rev. Econ. Sociol. Rural.* **2017**, *54*, 635–652. [CrossRef]
27. RIMISP. *Juventud Rural y Territorio*; Pobreza y Desigualdad Informe Latinoamericano: Santiago, Chile, 2019.
28. Rojas, F.; Rodríguez, P.; Marchant, C.; Troncoso, R. Los espacios rurales en Chile. Reflexiones sobre sus transformaciones e implicancias en las últimas cuatro décadas. In *Chile Cambiando: Revisitando la geografía Regional de Wolfgang Weischet*; Borsdorf, A., Marchant, C., Rovira, A., Sanchez, R., Eds.; Serie GEOlibros: Santiago, Chile, 2020; Volume 1, pp. 621–679.
29. Núcleo Milenio Centro para el Impacto Socioeconómico de las Políticas Ambientales (CESIEP). *Estudio de Indicadores de Calidad de Vida y Estándares de Vida en los Territorios Rurales de Chile*; ODEPA: Santiago, Chile, 2019.
30. Instituto Nacional de Estadística (INE). *Estimaciones y Proyecciones a Nivel Regional de la Población de Chile 2002–2035*; INE: Santiago, Chile, 2020.
31. Instituto Nacional de Estadística (INE). Estimaciones y Proyecciones de la Población de Chile 2002–2035. In *Totales Regionales, Población Urbana Y Rural. Síntesis de Resultados*; INE: Santiago, Chile, 2019.
32. Instituto Nacional de Estadística (INE). Redatam Diseminación de Datos. Available online: https://redatam-ine.ine.cl/ (accessed on 14 May 2021).
33. Abramovay, J.; Silvestre, M.; Cortina, N.; Baldissera, I.; Ferrari, D.; Testa, V. Sucessao proffissional e transferencia hereditaria na agricultura familiar. In Proceedings of the X Congreso Mundial de Sociología Rural, Río de Janeiro, Brazil, 2 August 2000.
34. Caputo, L. *Identidades Trastocadas de la Juventud Rural en Contexto de exclusión. Ensayando una reflexión Sobre la Juventud Campesina Paraguaya*; Documento de Trabajo no. 102; CLACSO: Asuncion, Paraguay, 2001.
35. Weisheimer, N. Os jovens agricultores e o processo de trabalho da agricultura familiar. In Proceedings of the VI Congreso de la Asociación Latinoamericana de Sociología, Porto Alegre, Brazil, 27 November 2002.
36. Kessler, G. La investigación social sobre juventud rural en América Latina. Estado de la cuestión de un campo en conformación. *Rev. Colomb. Educ.* **2006**, *51*, 16–39. [CrossRef]
37. Pardo, R. *Dianóstico de la Juventud Rural en Colombia. Grupos de Diálogo Rural, una Estrategia de Incidencia*; Serie documento N°227. Grupo de Trabajo Inclusión Social y Desarrollo. Programa Jóvenes Rurales, Territorios y Oportunidades: Una estrategia de diálogos de políticas; Rimisp: Santiago, Chile, 2017.
38. Díaz, V.; Fernández, J. *¿Qué Sabemos de los Jóvenes Rurales? Síntesis de la Situación los Jóvenes Rurales en Colombia, Ecuador, México y Perú*; Serie documento de Trabajo N° 228, Grupo de Trabajo Inclusión Social y Desarrollo. Programa Jóvenes Rurales, Territorios y Oportunidades: Una estrategia de diálogos de políticas; Rimisp: Santiago, Chile, 2017.
39. Stockdale, A. Migration: Pre-requisite for rural economic regeneration? *J. Rural Stud.* **2006**, *22*, 354–366. [CrossRef]

40. Garasky, S. Where are they going? A comparison of urban and rural youths' locational choices after leaving the parental home. *Soc. Sci. Res.* **2002**, *31*, 409–431. [CrossRef]
41. Bednaříková, Z.; Bavorová, M.; Ponkina, E.V.P. Migration motivation of agriculturally educated rural youth: The case of Russian Siberia. *J. Rural Stud.* **2016**, *45*, 99–111. [CrossRef]
42. Pelzom, T.; Katel, O. Youth Perception of Agriculture and potential for employment in the context of rural development in Bhutan. *Dev. Environ. Foresight* **2017**, *3*, 2336–6621.
43. Anton, C.; Lawrence, C. Home is where the heart is: The effect of place of residence on place attachment and community participation. *J. Environ. Psychol.* **2014**, *40*, 451–461. [CrossRef]
44. Alston, A. 'You don't want to be a check-out chick all your life': The out-migration of young people from Australia's small rural towns. *Aust. J. Soc.* **2004**, *39*, 299–313. [CrossRef]
45. Tonts, M. Competitive sport and social capital in rural Australia. *J. Rural Stud.* **2005**, *21*, 137–149. [CrossRef]
46. Chambers, R.; Conway, G. *Sustainable Rural Livelihoods: Practical Concepts for the 21st Century*; Institute of Development Studies: Brighton, UK, 1991.
47. Shucksmith, M. Young People and Social Exclusion in Rural Areas. *Sociol Rural.* **2004**, *44*, 43–59. [CrossRef]
48. Pérez, G. *Caminos Rurales: Vías Claves Para la Producción, la Conectividad y el Desarrollo Territorial*; Boletín FAL N°377; CEPAL: Santiago, Chile, 2020; Available online: https://hdl.handle.net/11362/45781 (accessed on 23 May 2021).
49. Phillipson, J.; Gorton, M.; Turner, R.; Shucksmith, M.; Aitken-McDermott, K.; Areal, F.; Cowie, P.; Hubbard, C.; Maioli, S.; McAreavey, R.; et al. The COVID-19 Pandemic and Its Implications for Rural Economies. *Sustainability* **2020**, *12*, 3973. [CrossRef]
50. Kirkpatrick Johnson, M.; Elder, G.H.; Stern, M. Attachments to family and community and the young adult transition of rural youth. *J. Res. Adolesc.* **2005**, *15*, 99–125. [CrossRef]
51. Brügger, A.; Kaiser, F.G.; Roczen, N. One for all?: Connectedness to nature, inclusion of nature, environmental identity, and implicit association with nature. *Eur. Psychol.* **2011**, *16*, 324–333. [CrossRef]
52. Wiborg, A. Place, nature and migration: Student's attachment to their rural home places. *Sociol. Rural.* **2004**, *44*, 416–432. [CrossRef]
53. Haukanes, H. Belonging, Mobility and the Future: Representations of Space in the Life Narratives of Young Rural Czechs. *Young* **2013**, *21*, 193–210. [CrossRef]
54. Ministerio del Interior y Seguridad Pública. *Política Nacional de Desarrollo Rural*; Diario Oficial: Santiago, Chile, 2020.
55. Godoy, M.; Contreras, H. Continuidad y cambio en una comunidad indígena del Norte Chico: Valle Hermoso, 1650–1950. In *Experiencias de Historia Regional en Chile: Tendencias historiográficas Actuales*; Cáceres, J., Ed.; Instituto de Historia PUCV: Valparaíso, Chile, 2008; pp. 219–237.
56. Moyano, C. *Oficios Campesinos del Valle de Aconcagua*; Ediciones Inubicalistas: Valparaíso, Chile, 2014.
57. Razeto Migliaro, J.; Catalán Martina, E.; Skewes Vodanovic, J.C. Soberanía territorial, conservación ambiental y comunidades de campo común en Chile central. *Polis Rev. Latam.* **2019**, *54*, 75–89. [CrossRef]
58. Bartlett, J.E., II; Kotrlik, J.W.; Higgins, C.C. Organizational Research: Determining appropriate sample size in survey research. *Inf. Technol. Learn. Perform. J.* **2001**, *19*, 43–50.
59. Almuna, R.; Cruz, J.M.; Vargas, F.H.; Ibarra, J.T. Landscapes of coexistence: Generating predictive risk models to mitigate human-raptor conflicts in forest socio-ecosystems. *Biol. Conserv.* **2020**, *251*, 108795. [CrossRef]
60. R Core Team. *R: A Language and Environment for Statistical Computing*; R Foundation for Statistical Computing: Vienna, Austria, 2021.
61. Revelle, W. *Psych: Procedures for Psychological, Psychometric, and Personality Research*; R Package Version 2.1.6; Northwestern University: Evanston, IL, USA, 2021.
62. Göb, R.; McCollin, C.; Ramalhoto, M.F. Ordinal methodology in the analysis of likert scales. *Qual. Quant.* **2007**, *41*, 601–626. [CrossRef]
63. Gardner, H.J.; Martin, M.A. Analyzing ordinal scales in studies of virtual environments: Likert or lump it! *Presence Teleoperators Virtual Environ.* **2007**, *16*, 439–446. [CrossRef]
64. Koenker, R.; Chernozhukov, V.; He, X.; Peng, L. (Eds.) *Handbook of Quantile Regression*; CRC Press, Taylor & Francis Group: Boca Raton, FL, USA, 2017. [CrossRef]
65. Koenker, R.; Hallock, K.F. Quantile regression. *J. Econ. Perspect.* **2001**, *15*, 143–156. [CrossRef]
66. Hummon, D. Community Attachment: Local Sentiment and Sense of Place. In *Place Attachment*; Altman, I., Low, S., Eds.; Plenum Press: New York, NY, USA, 1992; pp. 253–278.
67. Von Reichert, C. Returning and New Montana Migrants: Socio-economic and Motivational Differences. *Growth Chang.* **2001**, *32*, 447–465. [CrossRef]
68. Wang, W.W.; Fan, C.C. Success or failure: Selectivity and reasons of return migration in Sichuan and Anhui, China. *Environ. Plan. A* **2006**, *38*, 939–958. [CrossRef]
69. Niedomysl, T.; Amcoff, J. Why Return Migrants Return: Survey Evidence on Motives for Internal Return Migration in Sweden. *Popul. Space Place* **2011**, *17*, 656–673. [CrossRef]
70. Haartsen, T.; Thissen, F. The success-failure dichotomy revisited: Young adults' motives to return to their rural home region. *Child. Geogr.* **2014**, *12*, 87–101. [CrossRef]

71. Jurado, C.; Tobasura, I. Dilema de la juventud en territorios rurales de Colombia: ¿campo o ciudad? *Rev. Latinoam. Cienc. Soc. Niñez Juv.* **2012**, *10*, 63–77.
72. Kearney, A.R. Residential development patterns and neighborhood satisfaction: Impacts of density and nearby nature. *Environ. Behav.* **2006**, *38*, 112–139. [CrossRef]
73. Fränkel, S.; Sellmann-Risse, D.; Basten, M. Fourth graders' connectedness to nature—Does cultural background matter? *J. Environ. Psychol.* **2019**, *66*, 101347. [CrossRef]
74. Duron-Ramos, M.F.; Collado, S.; García-Vázquez, F.I.; Bello-Echeverria, M. The Role of Urban/Rural Environments on Mexican Children's Connection to Nature and Pro-environmental Behavior. *Front. Psychol.* **2020**, *11*, 514. [CrossRef]
75. Soga, M.; Gaston, K.J. Extinction of experience: The loss of human-nature interactions. *Front. Ecol. Environ.* **2016**, *14*, 94–101. [CrossRef]
76. Gosling, E.; Williams, K.J.H. Connectedness to nature, place attachment and conservation behaviour: Testing connectedness theory among farmers. *J. Environ. Psychol.* **2010**, *30*, 298–304. [CrossRef]
77. Basu, M.; Hashimoto, S.; Dasgupta, R. The mediating role of place attachment between nature connectedness and human well-being: Perspectives from Japan. *Sustain. Sci.* **2020**, *15*, 849–862. [CrossRef]
78. Larson, L.R.; Szczytko, R.; Bowers, E.P.; Stephens, L.E.; Stevenson, K.T.; Floyd, M.F. Outdoor Time, Screen Time, and Connection to Nature: Troubling Trends Among Rural Youth? *Environ. Behav.* **2019**, *51*, 966–991. [CrossRef]
79. Pergams, O.R.W.; Zaradic, P.A. Is love of nature in the US becoming love of electronic media? 16-year downtrend in national park visits explained by watching movies, playing video games, internet use, and oil prices. *J. Environ. Manag.* **2006**, *80*, 387–393. [CrossRef]
80. Ruiz Pulpón, Á.R.; Cañizares Ruiz, M.D.C. Potential of vineyard landscapes for sustainable tourism. *Geoscience* **2019**, *9*, 472. [CrossRef]
81. Santoro, A.; Venturi, M.; Agnoletti, M. Agricultural heritage systems and landscape perception among tourists. The case of Lamole, Chianti (Italy). *Sustainability* **2020**, *12*, 3509. [CrossRef]
82. Cortés, M.E.C. Sequía, degradación ambiental, trabajo y educación: Un breve comentario sobre la realidad actual de las comunidades agrícolas de la provincia del Limarí Chile. *Idesia* **2016**, *34*, 73–76. [CrossRef]
83. Ruíz Peyré, F. ¿Nacer en el campo-morir en la ciudad?: Exclusión y expulsión de los jóvenes de áreas rurales en América Latina. *Rev. Electr. Teoría Educ.* **2008**, *9*, 181–195. [CrossRef]

MDPI

Article

Sustainability and Agricultural Regeneration in Hungarian Agriculture

Imre Kovách [1,2], Boldizsár Gergely Megyesi [2,*], Attila Bai [3] and Péter Balogh [4]

1 Department of Sociology and Social Policy, University of Debrecen, H-4032 Debrecen, Hungary; Kovach.Imre@tk.hu
2 Institute for Sociology, Centre for Social Sciences, H-1097 Budapest, Hungary
3 Institute of Applied Economics, University of Debrecen, H-4032 Debrecen, Hungary; bai.attila@econ.unideb.hu
4 Department of Statistics and Methodology, University of Debrecen, H-4032 Debrecen, Hungary; balogh.peter@econ.unideb.hu
* Correspondence: Megyesi.Boldizsar@tk.mta.hu

Abstract: Generational renewal is a core issue in European agriculture. Despite the continuous efforts of governments and the EU Council, the ageing of farmers seems an unstoppable process, accompanied by land concentration, the decrease in agricultural activity and the transformation of the European countryside. Consequently, there is a very rich scientific literature analysing the problem; a great part of it argues that the young farmer problem consists, in fact, in a number of different problems, with these problems showing huge regional differences. Hungary, as a new member state, with a heterogeneous (both fragmented and concentrated) land-use structure offers a good field to analyse generational renewal. Our paper is based on the first results of an ongoing Horizon 2020 project analysing rural regeneration. As a part of the research study, 48 semi-structured interviews were conducted with young farmers, successors of farmers and new entrants into farming. In our paper, we explore how education, access to land and family traditions influenced generational renewal and how it impacts sustainability practices.

Keywords: generational renewal; sustainability; education; Hungary; access to land; farming traditions

Citation: Kovách, I.; Megyesi, B.G.; Bai, A.; Balogh, P. Sustainability and Agricultural Regeneration in Hungarian Agriculture. *Sustainability* **2022**, *14*, 969. https://doi.org/10.3390/su14020969

Academic Editor: John McDonagh

Received: 22 November 2021
Accepted: 12 January 2022
Published: 15 January 2022

Publisher's Note: MDPI stays neutral with regard to jurisdictional claims in published maps and institutional affiliations.

1. Introduction

It is clear, from publications on the world's food prospects, that young people's unwillingness to work in agriculture and large-scale exits from the sector appear to be accelerating and this is fundamentally contributing to the strategic challenges to be addressed in food regimes [1]. The world of working farms is also changing at a rapid pace. The traditional peasantry is permanently disappearing from the European countryside [2] and is being replaced by a variety of farmers, of which an EU research study lists ten types [3]. The comprehensive reports on sustainability and precision agriculture emphasize that, without supporting for generational renewal, the necessary restructuring of agriculture will not take place [4,5]. In Europe, 11% of farmers were under the age of 40 [6], the challenges of whom have rightly been highlighted by researchers [7–9].

The aging of the agricultural population is not typical in Hungary [10], but can also be considered common in developed economies. The exit of young people from agriculture has particularly detrimental consequences in regions of major importance for agricultural production [11]. In the 2000s, the proportion of Hungarian farmers under the age of 35 was estimated at around 20% and their utilized agricultural area was 12%. At the turn of the millennium, there were three times as many 65-year-old farmers as under-35 farmers. In 2010, this ratio rose to four and, in 2013, to nearly five. According to 2015 data, Hungary showed a similar picture to the EU, where the proportion of farmers under the age of 35 was also low, at 6.1% [6]. The number of recipients of young farmer support payments increased from 10,031 in 2016 to 12,722, but the amount of the payment decreased slightly, which indicates that new entrants to the support system used a smaller area [12]. Table 1 shows the number of farmers aged between 15 and 39 (eligible for young farmers' support) and their share in the whole generation. Both indicators reflect a smaller increase in the number of younger farmers, which, in turn, does not fundamentally change the age structure of the agricultural population.

Table 1. The number of young farmers eligible for support in 2017 (thousand people, %).

	2013	2014	2015	2016	2017
Famers between 15 and 39 years (thousand people)	66.9	68.6	74.6	78.4	77.2
Proportion of total number of farmers (%)	3.7	3.7	4.0	4.2	4.2

Source: [12], calculation based on EUROSTAT.

The average area of land cultivated by young farmers exceeds the national average, yet the basic development goal is to increase the size of young people's holdings. Through public auctions conducted within the framework of the "Land for Farmers" program, every third farmer—about 30,000 farmers—acquired land ownership at market prices, including more than 1200 young farmers. The land could be purchased for almost double the average price of national arable land in 2015 [13]. Between 2015 and 2016, more than 1200 young people under the age of 40 became the owners of 50,000 hectares of land under the "Land for Farmers" government program [14].

The only organization established in Hungary specifically for the benefit of young agricultural entrepreneurs is the Hungarian Association of Young Farmers (AGRYA). The association was founded in 1996 and currently counts more than 3000 young farmers (in the 25–35-year age group). However, even young farmers over the age of 35 do not necessarily have to leave the organization. Through the "Senior College", farmers older than 35 years can still be a part of the life of the organization. The association has a program called the Second Wave, which was started for farm children and young adults between the ages of 18 and 25 who were not yet engaged in farming on their own.

2. Agricultural Restructuring in Hungary

In this part, we analyse how Hungarian agriculture changed in the last two decades, after the EU accession in 2004. The analysis is based on statistical data and on existing literature. Rural and agricultural restructuring are inseparable in the Hungarian context, and are influenced by the subsidy system of the Common Agricultural Policy [15–18]; the economic processes, namely, the increase in the food process and the lack of available free workforce; and the policy of the Hungarian government, which aims to redistribute land and favours large-scale arable crop farming [19].

A mixed farming structure (huge companies (sometimes former state farms and cooperatives); medium-sized, usually family-based, farms; and a lot of small-scale farms (part-time and subsistence farms)) characterizes Hungarian agriculture [20]. The profound transformation of agriculture started in the first half of the nineties, when the socialist-type cooperatives were transformed into private companies, new-type cooperatives, or went bankrupt [21], and was finished by the time of EU accession [22–24].

In Hungary, land ownership and land use are split. Despite land ownership, land use is less fragmented and the number of farms is continuously decreasing [19]. Statistics as well as data on farm subsidies show a rapid and continuous decrease in the number of farms. In 2010, the number of farms was 351,000, which dropped to 235,000 by 2020 [25]. In earlier papers, we have presented the decrease in the number of farms and subsistence farms; we have also shown that, especially in the case of smaller farms, it means a simplification of farming [26].

As a result of the decrease in the numbers of farm units, the data show a concentration of land use and land property [19,26]. The dominance of medium-sized and bigger estates, the liberalization of the land market, the EU accession, the subsidies from the Common Agricultural Policy and the increasing investments into the sector resulted in the consolidation of agriculture. Of the 190,000 farms receiving EU support (SAPS), around 12,000 family farms and an additional 3000 companies dominate the Hungarian agricultural sector. The prices have risen and become stable. However, these processes had a negative effect on crop structure; monocultures are more wide-spread, cereal and maize production are the most common crops and animal husbandry is declining [27]. In addition, the entrance into farming became more difficult, as recent studies have shown [28].

In 2014, 0.8% of farms used one-third of the total arable land and 7.5% of farms cultivated 75% of the agricultural area [19]. Parallel to land concentration, agricultural employment decreased from around 1 million in 1988 to less than 350,000 in 2010 and only 20% of them were younger than 40 years old.

3. Theoretical Background

The literature on generational renewal is wide, even if we only concentrate on the most recent results of the studies analysing the case of Europe. Zagata and Sutherland [9], in their seminal paper, argued that the young farmer problem is not one problem, but a mixture of several, regionally different problems, also related to the differences among farm types, namely, the difference between farm successors and new entrants. The authors argued that countries with predominantly small farms are more likely to face the problem of aging of farmers, while countries with a less fragmented farm structure have more young sole-holders; they also emphasized that new member states are more frequently facing the problems of generational renewal.

Coopmans et al. [29], based on an EU research study, identified three conceptual phases and fourteen factors which help to understand farm generational renewal; the authors argued that generational renewal has a psychological element (successor identity, as it is called by the authors); an institutional element, called the farm succession process, comprising managerial, practical, legal and symbolic actions in order to transfer the farm; and, finally, they emphasized the necessity of farm development, i.e., the long process of changing the organizational structure and the strategy of the farm [29]. According to the authors, the phases of generational transfer are dependent on several external and internal factors. Using a sample of around 85 farms in several EU countries, the authors identified fourteen factors, grouped in four "spheres of influence", analysed in which phase the decisions are made and how the different factors influence the decisions of the farmers in generational renewal in the different phases [29]. The results of the comparative study show that agricultural policy has the main focus on increasing entry into farming. The authors argued that further research would be necessary to understand the role of Young Farmer Payment in farm succession.

Mann focused his investigation on a special group of farm successors, on the age group of 14–34-year-old farmers [30,31]. Based on a sample of Swiss farmers, he argued that female respondents tended to emphasize identity elements, such as continuing family traditions, or preference to work outdoors, while, among males, identity elements were more important at a younger age. Older male respondents took into consideration economic factors more often [30]. As several studies emphasized, in several cases, farm successors are hesitant as the income of farming is below the income of other economic activities. Mann argued that generational renewal is dependent on age [30]. Generational renewal is also dependent on retirement decisions, as Conway et al. emphasized [32]. They pointed to the necessity of encouraging early retirement to encourage intergenerational farm transfer.

According to an earlier research study conducted in Hungary in 2015 [33], 69.2% of the farmers, or a relative of the farmer, had farming traditions. The same study revealed that there is a relationship between farm size and farming family traditions; in the case of smaller farms, it is more likely that the farmer is from a farmer family, while, in the case of bigger farms, it is less likely.

The relevance of family traditions can be grasped in two facts; there are three main forms of acquiring the land owned and used by the farms, buying, inheriting and through restitution [33]. Buying is more typical in the case of farms using more than 100 hectares, inheritance is typical in farms below 20 hectares. According to the same study, the typical way to become a farmer is to start subsistence, then semi-subsistence farming and, after this preparatory phase, starting market-led farming activity [33].

If we analyse the source of agricultural knowledge, we see that a vast majority has acquired farming knowledge through everyday farming practices, specifically, 47.6% through everyday practices, whilst only 27.7% said that his or her knowledge stemmed from educational institutions. If we analyse the connection between the source of agricultural knowledge and family traditions in farming, we see that the role of learning from the practice is higher if there is a strong family tradition; it also means that there is a group of farmers which has a strong scholarly, professional agricultural knowledge, which can supplement family traditions.

In our paper, we use the most common definition of sustainability, which can be found in the Our Common Future report [34]; we understand it and its three dimensions as a platform idea [35], keeping in mind that it has a continuously changing understanding by the different actors.

We argue that focusing on a crucial point of generational renewal, on farm transmission, we could better understand the role of educational levels, family traditions and access to land in the process. We also would like to understand whether a move toward more sustainable practices can be detected among the members of the younger generation [10].

4. Methods and Research Questions

The main aim of the paper is to present agricultural regeneration in Hungary, its main motivators and to analyse how it influences the further development and spread of sustainable agricultural practices.

Based on the results of an ongoing Horizon 2020 research project (Ruralization: The opening of rural areas to renew rural generations, jobs and farms (GA 817642)), we aim to describe the patterns of farm succession on a sample of north-eastern Hungarian farmers, focusing on the momentum of farm transmission. We aimed at collecting the different patterns of farm succession and to understand the relationship among the factors influencing the decision of the farmers. We paid special attention to generational raptures or continuity and to the role of family traditions. We also involved, in the analysis, the role of education in agricultural renewal and the role of attitudes toward nature and toward working close to nature. By understanding these attitudes, we explore how sustainability appears in the narratives of the younger generation of farmers, how they understand it and how they translate it into their agricultural practices.

We conducted 48 semi-structured interviews [36] with young land successors and new entrants into farming under 40 years of age; these interviews constituted the basis of our analysis, but, as we show in the following sections, we used other interviews from three different research projects. There were no representative data on successors and new entrants into agriculture; therefore, we used snow-ball sampling to select our interviewees. The interviews covered a wide range of topics. We asked the interviewees about their family background, about their education and about their farms. We obtained detailed information about their farming practices, about their relationship with the previous owner of the farm and about their future plans. We used a context analysis to explore the role of family traditions, education, emotions and access to land in generational renewal and to understand the attitudes toward sustainability.

The site of the fieldwork was two counties in eastern Hungary, Hajdú-Bihar and Szabolcs-Szatmár-Bereg. In both counties, the proportion of the agricultural sector is traditionally high. Although industrialization has taken place since the mid-20th century, overall income was lower, while agricultural employment and income were significantly higher than the national average. In Hajdu-Bihar, the rural population lived more in small towns, but the proportion of small villages was higher in the southern micro-regions. The rurality in Szabolcs-Szatmár-Bereg was the world of small villages, which was supplemented by the former market towns. The family farm was the most important form of agriculture in terms of numbers. Land use was segmented in Hajdú-Bihar; in addition to the strong concentration, there were also many smaller farms. High-quality arable land was dominated by crop production (maize and cereals). There were also larger estates in Szabolcs-Szatmár Bereg county, but there was a much higher proportion of small farms. The cereal production here was complemented by vegetable and, especially, fruit production. Animal husbandry tended to take place on large farms in both counties, while, in line with national trends, meat and egg production was slowly being pushed out of small farms. Our respondents were exclusively from family farms, inheriting a conventional medium-sized farm. Most of them had a BA or MA in agricultural engineering. Some of them worked together with other family members. There were only a few female respondents, mirroring that the rate of women among farmers less than 40 years old was only 26%.

As part of the fieldwork of the EU Ruralization project, 21 interviews were conducted with newcomers to the countryside, 11 of which were with farmers producing specialized products. During the research study on precision farming (TKP2020-KKK-04; implemented with the support provided from the National Research, Development and Innovation Fund of Hungary, financed under the 2020-4.1.1-TKP2020 funding scheme), 30 in-depth interviews were conducted with farmers and experts and, in the MILAB project (research study supported by the Ministry of Innovation and Technology NRDI Office within the framework of the Artificial Intelligence National Laboratory Program), another 30 in-depth interviews were conducted with precision farmers, a third of whom were young.

5. Factors Influencing Generational Renewal—The Empirical Results

The analysis of our extremely rich empirical material was organized around three main topics, developed using the theoretical background presented above. First, we analysed how family and traditions influenced young farmers' decisions on continuing or starting a farm business; secondly, we analysed access to land; then, we continued with the role of education in these decisions. We analysed the attitudes toward sustainability in the case of each topic.

5.1. Family

The family background (material, financial funds, knowledge capital and mental attachment) was crucial in the motivations of young farmers, although older generations did not always encourage entry into farming. Secondary and tertiary education, which is already a consequence of family social capital, was an additional motivator. Barriers to young people's access to employment and income and the continuation of family farming were the main external motivating factor. However, farming had high social value based on continuity, similar to the old EU Member States [37], despite the fact that, in Hungary, agriculture operated in a collectivized form between 1960 and 1990.

> "I gained professional knowledge from my grandfather and father and from experience. Because we also study at university, but what we experience at home is the real value" (male, 34; BA).

According to our interviewees, family was especially important in the decision of young farmers to continue or leave agriculture. Usually, family members supported them to continue farming. Family was also one of the main sources of knowledge; below, we present that, obviously, not all knowledge types used by them stemmed from family members. Family remained a reference point for young farmers and the role of family was discussed intensively when presenting labour division within the family. It is also very important to mention that family was not necessarily presented as an arena of idyllic and smooth collaboration; young farmers, almost in all cases, had to argue, sometimes contradict, even struggle with elder family members to follow the innovative agricultural methods, or to modify farming strategies. In the following, we overview how family influenced farm generational renewal.

The support of the family to continue farming can be present in various ways and in various turning points of the lifecycle—in educational choices, as well as in moving back to the rural settlement after finishing university.

As an interviewee expressed,

> "I chose this occupation, because I have been always interested in what my father did, I planned to work with him later, after finishing the university. It was a common decision of the family and myself" (female, 35; MA).

However, it was also not rare that someone was following his or her grandparents in farming. The support of the family was important, as well as providing the necessary financial capital, land and management skills.

> "A friend of mine started farming as his father passed, so in a short time he sold the orchard, and simplified the farm, as he had no skills how to manage the workers, how to organize the farm; one cannot learn it in the school, he simply could not learn it from his parents, how it works" (male, 35; MA).

The family is also a source of knowledge; most frequently, two forms of knowledge stem from the family, knowledge about conventional agricultural practices, traditional agricultural knowledge (Reyes-Garcia 2014) and ecological knowledge [38]. The latter refers to knowledge about the ecological characteristics of a region, a sometimes underestimated source of knowledge in farming. Traditional agricultural knowledge still exists, but has less importance in the case of medium-sized farms. Knowledge about conventional agriculture practices was the most commonly used knowledge type by farmers in Hungary. Conventional agricultural practices and knowledge mean the modern methods of conventional agriculture. These methods became widespread in the 1970s and 1980s and were used by socialist-type cooperatives. These unsustainable practices mean high artificial nutrients and pesticide use and, usually, a simple crop structure. As we show in the following, the applied knowledge and practice was a source of conflict between the farmer and the heir.

> "My father hardly accepts, that we buy a new sowing machine, instead of using the old one; and after 2–3 years he also realized that now sowing costs less than 3–4 thousand forints (~EUR 10) for us, but if we would have to buy it as a service, it would be 12 thousand forints (EUR 45).
>
> But it is also difficult to explain that now after harvesting the sunflower, it is necessary to harrow the stems of the sunflower before ploughing. but I told him that he can say anything, I will do it, because it is not the same to work the half meter long pieces into the soil, or the smaller parts, so we mulched it, harrowed it, and then ploughed the plot; and it counts a lot"
>
> It is very typical that young farmers have to work together with their parents:
>
> "To be honest, in this area there are almost no young people who would farm for himself, most of them farms together with his daddy. Here, life starts at around 30–35, then one can start independent farming, so I do not know a real young farmer, who would be in his or her twenties and would really work on his own, without any assistance" (male, 34; BA).

Family traditions can appear also as a constraint. The members of the younger generations had to continue the family farm, even if they had different aspirations; as a young farmer explained,

> "To be honest I did not want to work in agriculture at all. Thus in 1997 my father got a heart attack he sold a lot of arable land, 40 hectares, then I was really young, around 17 years old, I lived my disco ages. And the agriculture was completely different, nothing was like now, so when he asked me, whether I would like to work in agriculture I said, no, you can sell the land, but as my Dad was ill

someone had to continue farming, and then I could not avoid starting it ... " (male, 41; MA).

There were also other constrains in inheriting family farming; these constrains appeared, often, as conflicts between the elder farmer, most frequently the father, and the heir, usually a son. The different conflicts can be understood as differences in farming methods, or the market position of the farm, but can be interpreted as differences in attitudes toward sustainability.

Obviously, there was a distribution of work in several cases, as the following quotation shows:

"I would say, that I have good work relationship with my parents, we can negotiate about the duties, although there are certain things which both my father and I am doing, but certain tasks are waiting for me; I am responsible for the paper work around subsidies and projects, but all these stuff: keeping contact with the offices, institutions, land issues are my job, while everyday farming issues are solved by my father" (male, 34; MA).

The solution whereby paperwork and subsidies are handled by the heir was a widespread solution according to our interview results, while strategic planning, if not conducted together, and tasks around agricultural production were the responsibility of the elder members of the family. As another interviewee reported about the subsidies,

"It must have been evolved this way, When we started in 2004 we could draw with hand on a piece of paper the areas we farmed, but since 2006 one has to provide a digital map, and I could do it, while for my father it would have been too difficult; he could have learnt it probably, but it was easier for me; so it developed this how" (male, 37; MA).

The beginnings of participation in family farming were motivated by the need to create one's own farm, but, in several cases, also by precarious job opportunities. Emergency decisions were also made.

"There were two options, one to sell everything and the other that I would take over the farm" (male, 32; BA).

However, the full management of family work, from administrative tasks to independent decision making, was gradually transferred to the young farmer over several years.

"I used to work for my father, then with my father, and then later my father worked for me" (male, 28; BA).

As a young farm successor said, farm transfer between generations was not always free of conflict.

Attitudes toward sustainability and sustainable practices seemed to be an important part of generational renewal and inheritance of farming. This is a topic in which the different knowledge forms used by the different family members may clash and the new farmer, the successor, can start building the farm according to his or her own ideas.

5.2. Access to Land

The state provides a maximum of 10 million (HUF) (around EUR 28,000) in the form of a tender for the purchase of land to young farmers, which is the price of 7–10 hectares of land and is not sufficient to start farming due to the tender restrictions. The purchase or lease of land is not supported in any other way by the state. Rather, it provides real support for those who need additional income to run their existing farm. Every year, 50,000–60,000 farms, mostly cultivating small areas, cease to exist in the Hungarian agricultural economy, but most of their land ends up in large estates. The privatization of the last vacant land began two years ago. Compared to 4.7 million hectares of cultivated land, the planting and ultimately privatization of 900,000 hectares of significant undivided jointly owned land is mandatory under state regulations, but the interviews showed that larger owners are preferred. As Table 2 below shows, young farmers had little access to land for arable crop production, which is the dominant sector of Hungarian agriculture; instead, their share of land use in intensive fruit production was higher.

Table 2. Farmers' age and cultivation sector, 2020 (%).

Cultivation Sector	14–39 Years Old	40–64 Years Old	65 Years and Older	Total
lawn	12	65	23	100
fruit	15	60	25	100
grape	11	64	25	100
arable land	12	61	27	100

Scheme 2021. Agricultural census 2020, preliminary results.

The integration of new farmers was particularly effective when they started farming during their (university) education. *"It was very interesting to know if I understood much of what we were doing at home or if I could contribute."* In market towns and villages where farming had a tradition, a good farmer had considerable prestige. The choice of the agricultural profession was not forced by the family in any of the cases. It was the decision of the successors and new entrant farmers, typically arising from the desires and experiences of childhood, working outdoors and love of nature.

> "I've been there for animals since I was a kid, so I've been moving around them since I could walk. Closeness is not my world, I like to be outdoors, to work outside … I'm out in the open, in nature, I don't think I need a better office. I think one of the most beautiful things about working with a living being, be it a plant or an animal" (male, 42; MA).

A very common situation was the distance between the place of residence and the location of the farm, so dual life was not uncommon. The young farmer couple was often forced to choose to either live in a small town, which is the place of farming, but the spouse has to commute to work and the children to school every day, or to live in a larger settlement in the area, from where the farmer has to leave every day to keep an eye on the site and the crop. The choice of a two-person life for the comfort of the family was, primarily, a characteristic of those with higher education.

> " … It is a very close to nature, very small idyllic settlement anyway, so I really like to be outside., Just don't have to stay there, don't have to sleep there. It is very pleasant anyway, it is excellent for rest and refreshment, it is excellent for escaping from the big city" (male, 32; BA).

Difficulties in accessing arable land could also cause the farm to be further away from the family home, just as the subjective values of the farmer could lead to the spatial separation of residence and farm.

> "I have a duality because I like the relatively big city, if we can say about Debrecen, the urban lifestyle, with the advantage that there are more opportunities to spend after work, but I like to be outside, in the nature, while working" (male, 34; Ph.D.).

A version of a two-person life (especially when the distance between the farm and the place of residence is large) is when the farmer lives out on the farm during the year and typically only moves home in the winter when the seasonal work is completed.

Part-time farmers had different income expectations and saw part-time work as a kind of "safety margin".

> " … Because I work, I don't see it as a full-time job. So overall, I don't look at the cost like I'm living off of it either—but I still leave it as an alternative that if I can improve by then, of course I want to do it full time" (male, 37; MA).

In almost all cases, farming started with the privatization of land following the change in regime in 1990, with the acquisition of property by parents and grandparents, which was the basis of the current size of the estate. There were three motivators for this, i.e., (1) the attachment to farming as a way of life, mostly in possession of previous cooperative employment and expertise; (2) investment purposes, supplemented by additional land purchases; and (3) in the absence of other employment opportunities, this was a secure self-employment opportunity in a context of high unemployment. The land of young farmers was mostly a legacy from the land privatization of the 1990s, which the family increased with additional purchases, sometimes through marriage or other succession.

It is a characteristic process of inheritance that the young person first acquires experience in the farms run by the parents and, at the same time, obtains a secondary or higher education in agriculture, then builds relationships as an employee of a large-scale farm. Then, he/she undertakes to set up his/her own farming, which may initially be linked to the agricultural activity of the parents or siblings. The purchase of new land by young farmers is limited by high land prices and the competitive advantage of farms of hundreds or thousands of hectares. High concentration of land use is the main obstacle to new generations entering farming. It was less common for the entire land of the parental farm to be passed on to the successor even in the life of the parents.

The spouses of successful young farmers, typically, did not take up full-time employment. Even the farmer himself was, in many cases, not working part-time to run his own farm. He/she was typically employed by a larger agricultural organization and even ran their own farm (over 50 or 100 hectares) part-time. This often resulted in a division of agricultural work, land use, machinery use and even ownership, which severely limited the spread of promising practices that could be followed. Inheritance has also begun on large farms of hundreds or thousands of acres; here, access to land does not limit the success of the young farmer. However, this example is not available for the majority.

We could register a clear positive attitude towards sustainable agriculture in the emotional motivations of young farmers, but not in all cases. As we show below, the positive emotional disposition did not necessarily result in sustainable practices. The frequent mention of the natural environment as an attractive factor in farming, the high level of childhood experiences, environmental awareness and knowledge, and the theoretical and practical commitment to alternative forms of farming all appeared in the interviews. The idea of self-sufficiency and sustainability emerged as a special motive. It was reported that they wanted to test their ability to provide for themselves individually or at the family level using only the resources at their disposal, but in circumstances that did not cause significant harm to their children.

> "After getting into town, I longed for it after 5–6 years and was always looking for an opportunity to do a job that could be done in a small town. And so he became interested in this topic, the idea of self-sufficiency, to become a little independent of the system" (male, 29; BA).

In certain cases, sustainability appeared as an aspiration, but it was contradiction with the ideas of proper farming.

> *"The crops have to be free of weeds, I like if there are no weeds, and the plants grow uniformly"*, as a young educated successor explained (female, 33; MA).

The concentration of land ownership and land use was a strong barrier to the access of new generations to arable land, but this can be a structural driver of sustainable development. The response of young farmers to the land shortage was, in many cases, a complete or partial shift to intensive production and alternative forms of farming, as Csurgó et al. (2016) and Megyesi [24] also argued.

5.3. Education

In Hungary, the number of students in agricultural higher education is growing, with about 1200–1300 people graduating every year. Experience has shown that students already have a high degree of career awareness in vocational education, with 2600–2700 students graduating each year. Statistics show that students whose families are engaged in farming on their own farm or where their parents have an agricultural background are more likely to remain in the sector. According to Kovách [19], one of the components of the structural change in Hungarian agriculture is the significant statistical correlation between the educational level of farmers and the size of the land used. Close to half (43%) of owners of farms over 200 acres have tertiary education. One-third of farmers overseeing between 100 and 200 hectares are graduates. Among farmers with over 100 hectares, there is virtually no primary education. The number of low-educated food producers is steadily declining and is relegated primarily to subsistence farms. The proportion of graduates in all farm size categories exceeds 21%. Slightly more than half of farmers have at least a high school education. According to the data of the Agricultural Census, the size of the arable land increases in direct proportion to the level of agricultural education for all age groups (Table 3).

Table 3. Agricultural education and average hectare of land in 2020.

Age Group (Years)	Agricultural Education	Hectares, Average
14–39	no	5.9
14–39	practical experience	8.3
14–39	basic level	20.3
14–39	medium level	31.8
14–39	high level	74.6
40–64	no	7.4
40–64	practical experience	9
40–64	basic level	21.9
40–64	medium level	38.5
40–64	high level	109
65 -	no	5
65 -	practical experience	7.7
65 -	basic level	15.8
65 -	medium level	31.1
65 -	high level	83.9

Source: [25] Agricultural Census 2020, preliminary results.

The majority of the interviewed young farmers had an agricultural degree, from a vocational high school diploma to a doctorate. Their age was between 22 and 37 years; they had typically 6–8 years of farming experience. However, this activity was already divided between full-time and part-time staff, the latter even being university lecturers. The advantage of higher education is the knowledge of the principle and technical possibilities of sustainable and extensive farming, the continuous strengthening of knowledge, network capital, project and application expertise.

An interviewee with three degrees in agricultural engineering, plant protection and phytology summarized the importance of the relationships as follows:

> "..and it feels good anyway, that when you get out of the university bench, everyone gets to work, gets here and there, and then in a few years, everyone, or mostly everyone, works in a profession. And then these former teammates run together within a work area, whether at events or whether one company is cooperating with another, or here, even if we think of the work of such area representatives, not necessarily working in extraction or at an acquiring company either. So, relationships are important, without which it wouldn't work. And that

if everyone makes connections, the immersion is bigger, so I think that's essential to a successful and sustainable farming" (male, 37; BA).

Young farmers had knowledge of project management and related information, the source of which was a comprehensive vocational and higher education, which was also facilitated by a network that had been developed during the high-school and university years and had been later maintained and expanded.

> " ... The university, however, formed a good foundation in this matter. From the point of view of education, from the point of view of the application, our application received an extra assessment in all respects, because ... . with a tertiary education, this was all positive in the field. ... . So we won this, the decision was made at the end of 2012 and from then on we had to work as a sole proprietor and that is when we started our activity here" (male, 41; MA).

Young farmers, according to our interviews, had some knowledge about sustainable practices and distinguished among them. We found that organic farming was not at all attractive to our interviewees, who mainly inherited a conventional farm and, as we argued above, learned the modern agricultural practices of the 1970s and 1980s from their parents.

> "We don't practice organic farming, not even reduced pesticide use; although I know that it would be an advantage in projects; but it worth the effort only, if one can ask for a higher price for the product at the end of the day" (male, 32; BA).

Agri-environmental practices were more widespread than organic farming; some of the farmers had long and positive experiences with them and they continued sustainable practices, as the following quotation shows:

> "We have agri-environmental contracts since 2004, of course it changed a lot in the last almost twenty years, but we learned the basic rules and follow the changes. Fertilization is based on soil analysis, we modify the quantity of fertilizers according to it, we have a more diverse crop structure, including leguminous plants, and use manure as well, which is very important, I think" (male, 32; BA).

In these cases, the knowledge of management practices was of great help for the farmers; thus, they were capable of engaging in sustainable practices.

6. Discussion and Concluding Remarks—Sustainability and Generational Renewal

We analysed the role of family, access to land and education in farm generational renewal and its impact on sustainability. Our analysis focused mainly on farm transfer, on conflicts and cooperation around the transfer. We found that a relatively easier access to land facilitated generational renewal and, although the different educational back-ground may have resulted in conflicts, it was still a motivational factor. Similarly to the assumptions of Coopmans et al. [29], subsidies did not play an important role in the case of the analysed group of farmers. Our analysis is in line with Mann's findings [30,31], indicating that working outdoors is an important factor for generational renewal; however, economic rationality was not neglected by our respondents.

We found that family and the personal attachment to farming as a way of life was a crucial issue for the young farmers to continue farming. It was also a constrain, whereby at least one member of the family had to work on the farm. There were three issues which softened this constrain, namely, emotions (attachment to nature), economic reasons and the again increasing prestige of farming. As we demonstrate, working outside in open-air areas, close to nature, was an important and highly valued element of farming according to our interviewees. The processes of Hungarian agriculture of the last two decades also made this decision easier; thanks to the EU accession, the markets became more secure and CAP subsidies ensured the profitability of farming. Thus, continuing with farming is an economically rationale decision. In addition, as a result of these favourable processes, agriculture regained, at least partially, its prestige in rural areas.

Access to land was perhaps the most crucial point for farmers, as our analysis showed, something that other studies have also emphasized [39,40]. After the EU accession, land became the scarcest resource in agriculture [19].

As earlier representative studies have shown, the educational background of the farmers still stems from practice and only a bit more than one-fourth gained agricultural education in formal institutions (vocational schools or universities) [33]. At the same time, statistical data show that the higher formal education a farmer has, the larger the farm is. The educational background seems to influence the process of farm transfer; highly educated heirs seem to undertake conflicts about agriculture contracts more frequently.

As in other studies, in this one, we found that farmers did not consider the pursuit of innovative and sustainable precision agriculture to be age-dependent. It is widely believed that greater experience and a wide network of contacts are needed for successful practices in sustainable precision and organic farming [41]. Despite the international literature [42,43] which attaches fundamental importance to age in the spread of precision farming, Hungarian research has only partially confirmed this [44]. According to the empirical analysis, age played a role only as a secondary indicator of knowledge; however, knowledge, land quality and costs also determined the transition to innovative precision farming to a lesser extent than production and technical utility.

In the analysis, we focused more deeply on the role of notions and ideas of sustainability in farm generational renewal. According to our interviews, sustainability was an important issue for the farmers, but their understanding of the concept was different from the understanding of the literature. We analysed, in our interviews, how sustainability appeared in them and found that sustainability is linked to the following:

- Agricultural practice;
- Economics of farming;
- Natural environment and environmental protection.

Sustainability was understood, by the members of the younger farmer generation, as an agricultural practice. Most commonly, it meant a reduced pesticide, as well as a reduced artificial nutrient use. However, apart from this, sustainable agricultural practice had a highly diverse meaning. Sustainable practices were justified by economic reasons; either by sparing money on input materials or by the possibility to access subsidies (for example agri-environmental subsidies) by using sustainable methods.

Another common understanding of sustainability could be traced back to economic viability; farming incomes had to cover farming costs and profit should be maximised, while environmental externalities were usually not considered.

The third and perhaps most common understanding of sustainability was related to nature, to the natural environment, nature conservation, or sometimes to environmental protection. It was also mixed, in several cases, with unsustainable agricultural practices. As we showed, some of the farmers were emotionally attached to nature; it was also a reason for being a farmer and, as they also highly appreciated tidy plots, these were linked to each other—although it was clear that, without herbicide use, there are no clean plots. For them, sustainability meant the preserving of the landscape, the well-known natural environment.

Generational renewal opens up space for sustainability transition in all the analysed topics, but, by opening up space for the discussion about sustainability, it would be possible to make the links between more obvious these topics and to strengthen the emergence of sustainability practices. In this process, both policy and education could play a major role.

Author Contributions: Conceptualization, I.K. and B.G.M.; methodology, I.K. and B.G.M.; software, I.K. and B.G.M.; validation, B.G.M.; formal analysis, I.K. and B.G.M.; investigation, A.B., P.B., I.K. and B.G.M.; resources, I.K.; data curation, B.G.M.; writing—original draft preparation, I.K. and B.G.M.; writing—review and editing, I.K. and B.G.M.; supervision, I.K.; funding acquisition, I.K. All authors have read and agreed to the published version of the manuscript.

Funding: This research was funded by European Commission: GA 817642, the National Research, Development and Innovation Office: 2020-4.1.1-TKP2020, the Ministry for Innovation and Technology: Artificial Intelligence National Laboratory Program and the NKFIH 132676. Boldizsár Gergely Megyesi has a Bolyai János Post-doctoral Stipendium (and Hungarian Academy of Sciences) and a New National Excellence Program ÚNKP 2021-5 Stipendium. The APC was funded by the Ruralization project (GA 817642) Horizon 2020 of the European Commission.

Institutional Review Board Statement: The study was conducted in accordance with the Declaration of Helsinki, and approved by the Institutional Review Board (or Ethics Committee) of Human Research Ethics Committee, TU Delft on the 31st of October 2019.

Informed Consent Statement: Informed consent was obtained from all subjects involved in the study.

Data Availability Statement: The data are available at the office of the Department for Sociology and Social Policy, University of Debrecen.

Conflicts of Interest: The authors declare no conflict of interest.

References

1. Food and Agriculture Organization of the United Nations (Ed.) *The Future of Food and Agriculture: Trends and Challenges;* Food and Agriculture Organization of the United Nations: Rome, Italy, 2017; ISBN 978-92-5-109551-5.
2. Granberg, L.; Kovách, I.; Tovey, H. (Eds.) *Europe's Green Ring;* Perspectives on Rural Policy and Planning; Ashgate: Aldershot, Hampshire, UK; Burlington, VT, USA, 2001; ISBN 978-0-7546-1754-9.
3. European Commission. Joint Research Centre. In *Farmers of the Future;* Publications Office: Brussels, Belgium, 2020.
4. Oberč, B.P.; Arroyo Schnell, A. *Approaches to Sustainable Agriculture: Exploring the Pathways towards the Future of Farming;* IUCN, International Union for Conservation of Nature (IUCN): Brussels, Belgium, 2020; ISBN 978-2-8317-2054-8.
5. European Parliament. Directorate General for Parliamentary Research Services. In *Precision Agriculture and the Future of Farming in Europe: Scientific Foresight Study;* Brussels, European Union Publications Office: Brussels, Belgium, 2016.
6. Farm Indicators by Agricultural Area, Type of Farm, Standard Output, Sex and Age of the Manager and NUTS 2 Regions—Products Datasets—Eurostat. Available online: https://ec.europa.eu/eurostat/web/products-datasets/-/ef_m_farmang (accessed on 22 November 2021).
7. May, D.; Arancibia, S.; Behrendt, K.; Adams, J. Preventing Young Farmers from Leaving the Farm: Investigating the Effectiveness of the Young Farmer Payment Using a Behavioural Approach. *Land Use Policy* **2019**, *82*, 317–327. [CrossRef]
8. Zagata, L.; Lošt'Ák, M.; Swain, N. Family Farm Succession of the First Post-Socialist Generation in the Czech Republic. *East. Eur. Countrys.* **2019**, *25*, 9–35. [CrossRef]
9. Zagata, L.; Sutherland, L.-A. Deconstructing the 'Young Farmer Problem in Europe': Towards a Research Agenda. *J. Rural. Stud.* **2015**, *38*, 39–51. [CrossRef]
10. Kőszegi, I.R. The importance of environmental sustainability among young farmers in the Homokhátság (partial results from field research)A környezeti fenntarthatóság fontosságának vizsgálata a homokháti fiatal gazdák körében (egy primer kutatás részeredményeinek ismertetése). *GAZDÁLKODÁS Sci. J. Agric. Econ.* **2019**, *63*, 40–57. [CrossRef]
11. Privóczki, Z.I. Fiatal gazdák árbevétel és jövedelem elvárásai a vidékfejlesztési program induló támogatásának tükrében. *Taylor Gazdálkodás-És Szerv. Folyóirat* **2019**, *10*, 93–100.
12. Mizik, T. A Közös Agrárpolitika 2013. Évi Közvetlen Támogatási Rendszerének Hatásai a Magyar Mezőgazdaságra. *Közgazdasági Szle* **2019**, *66*, 1210–1229. [CrossRef]
13. FM: Lezárulnak a Földárverések. Available online: https://www.agrotrend.hu/hireink/fm-lezarulnak-a-foldarveresek (accessed on 22 November 2021).
14. Kamara, N.A. Középpontban a Fiatal Gazdák. Available online: https://www.nak.hu/agazati-hirek/videkfejlesztes/161-gazdasagfejlesztes/97533-kozeppontban-a-fiatal-gazdak (accessed on 22 November 2021).
15. Gorlach, K.; Lošťák, M.; Mooney, P.H. Agriculture, Communities, and New Social Movements: East European Ruralities in the Process of Restructuring. *J. Rural. Stud.* **2008**, *24*, 161–171. [CrossRef]
16. Brown, D.L.; Kulcsár, L.J.; Kulcsár, L.; Obádovics, C. Post-Socialist Restructuring and Population Redistribution in Hungary*. *Rural. Sociol.* **2005**, *70*, 336–359. [CrossRef]
17. Swain, N. *Green Barons, Force-of-Circumstance Entrepreneurs, Impotent Mayors: Rural Change in the Early Years of Post-Socialist Capitalist Democracy;* CEU Press: Budapest, Hungary; New York, NY, USA, 2013; ISBN 978-615-5225-70-3.
18. Kovács, K. The Agricultural Restructuring in Hungary 1990–2001. *Geogr. Pol.* **2003**, *76*, 55–72.
19. Kovách, I. *Földek és emberek—Földhasználók és földhasználati módok Magyarországon;* DU Press: Budapest, Hungary, 2016; ISBN 978-963-8302-50-2.
20. Kelemen, E.; Megyesi, B. The Role of Collective Farmers Marketing Initiatives. *East. Eur. Countrys.* **2007**, *13*, 97–111.
21. Csite, A.; Kovách, I. Hatalom és társadalom. In *A posztkommunizmus vége;* Napvilág Kiadó: Budapest, Hungary, 2002.
22. Tisenkopfs, T.; Kovách, I.; Lošťák, M.; Šūmane, S. Rebuilding and Failing Collectivity: Specific Challenges for Collective Farmers Marketing Initiatives in Post-Socialist Countries. *IJSAF* **2011**, *18*, 70–88.
23. Kovách, I. *A vidék az ezredfordulón;* Argumentum Kiadó: Budapest, Hungary, 2012; ISBN 978-963-446-679-6.

24. Megyesi, B. Landscape after Accession: The Effects of Agricultural and Rural Policies on Farming—Results of Case Study Conducted in Western-Hungary. *Hétfa Work. Pap.* **2016**, *18*, 28.
25. Hungarian Central Statistical Office. Available online: https://www.ksh.hu/agricultural_census_fss (accessed on 22 November 2021).
26. Csurgó, B.; Kovách, I.; Megyesi, B. After a Long March: The Results of Two Decades of Rural Restructuring in Hungary. *East. Eur. Countrys.* **2018**, *24*, 81–109. [CrossRef]
27. Kovács, K. Structures of Agricultural Land Use in Central Europe. In *Reflecting Transformation in Post-Socialist Rural Areas*; Heinonen, M., Nikula, J., Kopoteva, I., Granberg, L., Eds.; Cambridge Scholars Publishing: Newcastle, UK, 2007; pp. 87–114.
28. Hamar, A.; Kovács, K.; Váradi, M.M. Azért kell a föld, hogy ha a fiam mezőgazdaságból akar élni, ne csak tehenész lehessen más telepén. In *Földből Élők: Polarizáció a Magyar Vidéken*; Argumentum Kiadó: Budapest, Hungary, 2016; pp. 370–394, ISBN 978-963-446-773-1.
29. Coopmans, I.; Dessein, J.; Accatino, F.; Antonioli, F.; Bertolozzi-Caredio, D.; Gavrilescu, C.; Gradziuk, P.; Manevska-Tasevska, G.; Meuwissen, M.; Peneva, M.; et al. Understanding Farm Generational Renewal and Its Influencing Factors in Europe. *J. Rural. Stud.* **2021**, *86*, 398–409. [CrossRef]
30. Mann, S. Understanding Farm Succession by the Objective Hermeneutics Method. *Sociol. Rural.* **2007**, *47*, 369–383. [CrossRef]
31. Mann, S. Tracing the Process of Becoming a Farm Successor on Swiss Family Farms. *Agric. Hum. Values* **2007**, *24*, 435–443. [CrossRef]
32. Conway, S.F.; McDonagh, J.; Farrell, M.; Kinsella, A. Going against the Grain: Unravelling the Habitus of Older Farmers to Help Facilitate Generational Renewal in Agriculture. *Sociol. Rural.* **2021**, *61*, 602–622. [CrossRef]
33. Csurgó, B.; Kovách, I.; Megyesi, G.B. Földhasználat, üzemtípusok, gazdálkodók. In *Földből Élők: Polarizáció a Magyar Vidéken*; Argumentum Kiadó: Budapest, Hungary, 2016; pp. 34–65, ISBN 978-963-446-773-1.
34. World Commission on Environment and Development (Ed.) *Our Common Future*; Oxford paperbacks; Oxford University Press: Oxford, UK; New York, NY, USA, 1987; ISBN 978-0-19-282080-8.
35. Bruckmeier, K.; Tovey, H. Knowledge in Sustainable Rural Development: From Forms of Knowledge to Knowledge Processes. *Sociol. Rural.* **2008**, *48*, 313–329. [CrossRef]
36. Kvale, S. Ten Standard Objections to Qualitative Research Interviews. *J. Phenomenol. Psychol.* **1994**, *25*, 147–173. [CrossRef]
37. Murtagh, A.; Farrell, M.; Mahon, M.; McDonagh, J.; Conway, T.; Conway, S.; Altes, W.K. *D3.1 Assessment Framework—Ruralization*; NUI: Galway, Ireland, 2020; p. 63.
38. Berkes, F.; Colding, J.; Folke, C. Rediscovery of Traditional Ecological Knowledge as Adaptive Management. *Ecol. Appl.* **2000**, *10*, 1251–1262. [CrossRef]
39. Gonda, N. Land Grabbing and the Making of an Authoritarian Populist Regime in Hungary. *J. Peasant. Stud.* **2019**, *46*, 606–625. [CrossRef]
40. Kovács, K. *Földből élők: Polarizáció a Magyar Vidéken*; Argumentum Kiadó: Budapest, Hungary, 2016; ISBN 978-963-446-773-1.
41. Balogh, P.; Bai, A.; Czibere, I.; Kovách, I.; Fodor, L.; Bujdos, Á.; Sulyok, D.; Gabnai, Z.; Birkner, Z. Economic and Social Barriers of Precision Farming in Hungary. *Agronomy* **2021**, *11*, 1112. [CrossRef]
42. Barnes, A.; Sutherland, L.-A.; Toma, L.; Matthews, K.; Thomson, S. The Effect of the Common Agricultural Policy Reforms on Intentions towards Food Production: Evidence from Livestock Farmers. *Land Use Policy* **2016**, *50*, 548–558. [CrossRef]
43. Daberkow, S.G.; McBride, W.D. Farm and Operator Characteristics Affecting the Awareness and Adoption of Precision Agriculture Technologies in the US. *Precis. Agric.* **2003**, *4*, 163–177. [CrossRef]
44. Balogh, P.; Bujdos, Á.; Czibere, I.; Fodor, L.; Gabnai, Z.; Kovách, I.; Nagy, J.; Bai, A. Main Motivational Factors of Farmers Adopting Precision Farming in Hungary. *Agronomy* **2020**, *10*, 610. [CrossRef]

Article

Sustainable Family Farming Futures: Exploring the Challenges of Family Farm Decision Making through an Emotional Lens of 'Belonging'

Lorraine A. Holloway [1,*], Gemma Catney [1], Aileen Stockdale [2,†] and Roy Nelson [3]

1 Geography, School of Natural and Built Environment, Queen's University Belfast, Belfast BT7 1NN, UK; g.catney@qub.ac.uk
2 Planning, School of Natural and Built Environment, Queen's University Belfast, Belfast BT7 1NN, UK
3 School of Biological Sciences, Queen's University Belfast, Belfast BT7 1NN, UK; r.nelson@qub.ac.uk
* Correspondence: lholloway02@qub.ac.uk
† Deceased 27 March 2021.

Abstract: This paper illustrates the importance of moving beyond an economic focus, and towards an emotional one, to gain a more comprehensive understanding of why farmers can be reluctant to retire and/or pass their farm onto the next generation. We report on a two-phase qualitative study of family farm decision-making processes in Northern Ireland, drawing on 62 in-depth oral life history interviews with farmers, farmers' spouses, and farm successors. In an attempt to gain a deeper understanding of the emotional aspects of retirement and succession decision-making processes, and their relationship with place belonging, in the first phase of this research we employed an innovative 'Work and Talk' method, whereby interviews were conducted while shadowing, or in some cases, co-working, with farmers on their land. The second phase of this research responded to restrictions arising from the COVID-19 pandemic, and involved remote telephone or online interviews with family farm members. This research revealed the complex relationships between a 'longing for belonging' and emotional attachment to the family farm, and the challenges associated with patrilineal farming structures, expectations and identities, in planning for succession. The emotional impacts of strained relationships with policy makers around support for retirement emerged as a surprisingly dominant theme throughout the interview process, suggesting the need for greater emphasis on the emotional aspects of farming retirement and succession planning to inform future rural development policies targeted towards the sustainability of family farms.

Keywords: family farm; ageing farmers; retirement; succession; emotions; decision making; belonging; respect; rural sustainability; Northern Ireland

Citation: Holloway, L.A.; Catney, G.; Stockdale, A.; Nelson, R. Sustainable Family Farming Futures: Exploring the Challenges of Family Farm Decision Making through an Emotional Lens of 'Belonging'. *Sustainability* **2021**, *13*, 12271. https://doi.org/10.3390/su132112271

Academic Editor: John McDonagh

Received: 28 September 2021
Accepted: 2 November 2021
Published: 6 November 2021

Publisher's Note: MDPI stays neutral with regard to jurisdictional claims in published maps and institutional affiliations.

1. Introduction

There are over 500 million farms across the Global North and South, which are mostly managed by farm families [1]. The EU and the UK are reliant on family farming for economic sustainability, where in Northern Ireland (NI) the family farm and the agri-food industry plays a more significant economic role than in the rest of the UK [2]. Yet, family farming in NI offers more than just economic advantages; it is commonly viewed as the 'heart of rural communities' [3] because of its positively perceived values, its endurance over time and between generations, and its cultural assets [4].

The family farm, and, more broadly, rural communities have transformed over the last 100 years due to technological advances, globalisation and demographic changes, where government intervention and EU policies have, in part, been successful, for example in improving quality of rural life and supporting farm income [5]. However, in NI, where family farming remains patrilineal, a recent survey by the Ulster Farmers' Union highlighted that there are persistent policy shortcomings concerning ageing farmers, particularly in relation to retirement decision making and issues of rural belonging [6].

To date, the agricultural and rural literatures have focused on farmers' economic 'roots' to place, yet there is less research on how modern rural communities can affect the emotional, temporal and spatial connection to place amongst farmers, such as attachment to the land or farm animals. Research in Australia has identified that a loss of a 'sense of belonging' can impact not only on farm decision making but also the health and well-being of farming families [7]. The economic focus of the effects of rural change on farmers' decision making fails to understand the emotional dynamics behind why farmers can be reluctant to let go of the family farm. As argued by Errington and Gasson, family farm members' emotional attachments to place can often supersede rational or economic judgements within the decision-making process [8].

This paper aims to utilise an emotional, rather than an economic, approach, to investigate how a 'sense of belonging' within changing rural communities can shape family farm retirement and succession decision-making processes in Northern Ireland. In recognition of the need for greater understandings of the effects of farmers' emotions [9], this research explores how farmers' embodied feelings (physical and emotional) alongside their experiences (memories) of belonging can affect their perceptions of, and anxieties around, retirement and succession. This research considers the family farm as both a 'place to belong' and a 'sense of belonging' [10]. The paper draws on interviews with 62 people across small livestock farms in Northern Ireland, with 21 farming men, 20 women spouses of farmers, and 21 adult children of farmers, of which all but one were successors. An innovative Work and Talk methodology was applied in an attempt for the interviewer to dig deeper in conversations with farming families. These involved, for example, shadowing and assisting participants on the farm or in the family home. Northern Ireland is an interesting case study in which to understand these patrilineal embodied emotions based on its agricultural reliance and emotive historical land ownership.

1.1. Emotions, the Family Farm and Decision Making

Emotions are a part of our human experience and they are not only essential to helping us describe the world around us, but also help us shape what we think of it and how we want to respond to it [11]. Emotions are difficult to define; however, in simple terms, they might be ultimately understood as our feelings and how we react to positive or negative situations, which are also 'corporeal', permitting us to comprehend how we feel and think [12]. Emotions, however, are not inherent; they can be acquired [9,13]. Anderson and Smith's much cited editorial explores the links between places, spaces and emotional experiences [14]. Paying attention to emotion in our research allows us to appreciate how people's lives are lived and experienced, and in turn how emotions affect their environment based on emotional bonds formed, which also shape one's sense of identity [14]. An example of this raw embodiment of thinking and feeling is illustrated in Pini et al.'s study of a coal mine closure in rural Australia, which demonstrates emotional loss through the feelings of betrayal, anger and resentment to the closure of the mine that dominate over any economic loss [15].

There is still much to learn from understanding the emotions of farmers and their families within changing rural communities. Notable exceptions include Ramirez-Ferrero's book 'Troubled Fields', which demonstrates that it is not just economic issues affecting the mortality rate of farming men in a time of crisis, but social and cultural issues because of rural modernisation [16]. Price and Conn's research on keeping the name on the land in Northern Ireland also identifies that succession is an emotional process, rather than a practical one [17]. Rieple and Snijdger's work recognises the need to understand the 'emotional dimensions' of family farm decision making; in their case, within traditional farming structures [9]. Undertaking semi-structured interviews with 27 dairy farmers in Munster (Republic of Ireland; RoI), Rieple and Snijdger explored emotions to innovation on the farm, concluding that decision making is based on various emotional factors, such as satisfaction with a traditional farming 'lifestyle', socio-emotional bonds within communities, safeguarding the continuity of the family farm, and a preference for traditional, rather than economic, decision making to cattle selection [9]. Conway et al.'s research in the RoI illustrates further that emotional issues, such as loss of farming identity, status, power and the relationship with the farm, especially in later life, are affecting older farmers' retirement decision making [18,19]. Their paper also argues that more evidence is needed to understand emotional attachment to the farm and its 'embodied contents', such as livestock, land, and the farm. Glover and Reay's study on dairy farms focuses on the non-financial socio-emotional wealth of farmers, illustrating that in order to encourage future farm sustainability, policy makers need to be aware that, regardless of any economic issues on the farm, the value, goals and emotional attachment of the farm family are key to farm survival, and that this is evidenced even by farmers without a successor [20]. As Grubbstrom and Erickson's research on retired farmers' decision making in Sweden illustrates, it is emotional incentives such as intrinsic values, care for the environment, land and rural community which entice them to either sell or lease their property to the next generation [21].

Place attachment(s) is important when trying to understand emotions on the family farm because it is argued that the more time spent in place, the greater the emotional attachment to place [22]. This is especially evident for older people who might naturally have spent a longer length of time in one place, building relationships, memories and bonds [23]. Yet, while elective belonging in urban space has received scholarly attention [24], there is a need for better understandings of belonging or elective belonging in rural areas, as proposed by Stockdale et al. (2018) [25]. Erickson et al.'s paper on rural stayers argues that research is still lacking on the importance of place and how it can enhance qualities such as community attachment, rootedness and a sense of belonging [26]. Stockdale and Ferguson's research in NI suggests that stayers in rural areas demonstrate a strong sense of place attachment and belonging, but that these are intertwined with complicated relationships with family history, farm ownership and continued family networks [27]. This emotional place attachment of 'being at home' is heightened further when relied on for 'survival' [28], such as farms, which have a clear socio-economic role.

1.2. The Emotion of Belonging

It is widely accepted that people share a desire to 'belong', and to have a sense of attachment [29,30]. However, it is only relatively recently that belonging within rural communities has been the subject of academic attention [31]. Belonging is multidimensional and can mean many things; it can refer to a place, identity, thoughts, and our emotions and feelings, which can have a direct reflection on our personal experience of belonging [10,14,29]. Allen et al., in defining belonging, argue that a need to "connect deeply with other people" is a common factor, but that direct involvement is not always necessary in order to connect, and instead can be based on the quality, experiences and observations of these connections [32] (p. 88). Others will gain on a 'sense of belonging' through their social connections, for example, through their childhood experiences, and place where they grew up [33].

To put this relative to family farming, and in particular to farming retirement and succession, because farms are commonly passed down through the generations, they offer a treasure chest of social and familial inter-relations and experiences that can impact on family members' sense of belonging [34]. For example, Price and Evans, in their case study of rural mid-Wales [35], identify the evolving relationship with farming women. Price and Evans reveal how, unless women were from a traditional farming background, they were often considered by farming men as a threat to the success and survival of the family farm, in respect to issues around marital breakdown and inheritance, ultimately affecting succession planning. Price and Evans also discuss how the complex patrilineal culture means that farming men tend to only seek emotional help from those that are from a common background, or farming 'way of life', and as rural communities change, this support culture is becoming at risk [35]. Yet, it is not just the inter-relationships between *people*, belonging and retirement decision making. Riley, for example, explores the roles of farmers' livestock in shaping farmers' identity and attachment, arguing that "animals are central to the everyday lives and identities of farmers ... separation from them alters farmers' attachment to particular practices, places and social networks" [36] (p. 2).

Place attachment—to the farm and farmland—forged as it may through relationships with family, community, animals, and through inheritance, history and memory, is a central aspect of farmers' identity and belonging. Riley identifies the importance of, and opportunities presented by, making 'place' central within the research process as it provides important insights into farming identities, emotions and social relationships [37]. An example of this is the traditional patrilineal farming structure, which provides a farmer with a sense of (place) identity, and one farmers can be reluctant to let go of, especially in retirement. Bryant and Pini noted that a sense of belonging and age are interlinked in rural communities, especially with older farming men, as seen through their 'pride of place', developed through the generations of family land ownership [31] (p. 136). Price and Conn identify this in their study by exploring the continuing reliance of the patrilineal requirement to keep the name on the land; they show how place attachment through the 'pull of the land' has a significant influence on succession planning in NI [17]. Price and Evans's ethnographic case study of farm families in Wales identified that attachment to a farming 'way of life', experienced through farming identity, relations, roles, and home, can contribute and/or cause farming 'distress' [35].

Our research objective is to gain a better understanding of the emotion of belonging in family farming, and how this influences the retirement decision-making processes. It might be argued that without this awareness, current rural development policies are paying insufficient attention to the importance of traditional patrilineal farming identities within changing rural communities, and the unique challenges experienced by farmers faced with losing, or at least weakening, their farming identity and place belonging, as they transition to retirement.

2. Northern Ireland Context

The family farm and the agri-food industry play a more significant economic role in Northern Ireland (NI) than in the rest of the UK, with an annual turnover overall of £4.5bn each year from over 24,827 active farms [38]. Approximately 75% of the total land area in NI (1.35 m hectares) is agricultural, albeit with a decline by 1% since 2019 [39]. Most farms in NI today are classed as 'very small' (76%), operated by families and supported by family labour with a reliance on the patrilineal family farm structure. Price and Conn's mixed methods research examining the prerequisites of 'keeping the name on the land' demonstrated the possibilities offered by an NI case study for understanding patrilineal retirement and succession decision making [17].

NI's agricultural space can be considered unique compared with the rest of the UK because of its shared border with an EU State through the RoI [40]. Economically, DEFRA has also identified an inadequate acknowledgment of the essential differences between the agricultural sector of NI and other parts of the UK [41]. Price and Simpson argue that it is important for legislation to acknowledge NI's 'otherness' to the rest of the UK because of its agricultural dependence and region-specific characteristics [42]. Indeed, NI's historic struggles with land ownership via political conflict represents a key difference to the rest of the UK [42], and land in NI, much like the RoI, is very much a part of NI's rural culture. Land, of course, is an asset and vital for most rural economies for growth in part, but for NI (and RoI), there is link between land and identity where it has acted to control belonging at micro levels and divide communities at macro levels [42]. Land in NI is also linked to freedom of land ownership, a unique land tenure system which defines how property rights are allocated, transferred, and used [43,44]. In relation to farming in NI, this includes a 'fee farm grant' which is comparable to that of a freehold title where land can be leased indefinitely and sold on in the same basis through the family to secure the title of land ownership for the tenant [43].

In short, NI's uniqueness, through its patrilineal agricultural reliance, culturally-motivated emotive relationship with land ownership, and land protection through its legislation [42], makes it an interesting case study to explore the emotional decision making of patrilineal farmers and their families.

3. Methodology

This research, undertaken on the family farm, intends to elucidate how farmers' embodied (physical and emotional) feelings alongside their experiences (such as memories) of belonging can shape their retirement and succession decision making. A Constructivist Grounded Theory approach was deemed appropriate for this research as it can be rewarding for studies where very little theory exists [45–47]. The collected data were reviewed as an ongoing process throughout the analytical phase, in order to allow core concepts around themes to emerge [45].

3.1. Research Sample

Several factors influenced the semi-structured interview sampling strategy. Farmers aged 50+ are most likely to be in the process of making retirement or succession decisions. In Northern Ireland, the average age of the farmer is 58, with only 6% of farmers under the age of 35 [48]. We therefore aimed to recruit farmers and successors between the ages of 18 and 75, with most farmers aged 50 and over. The agricultural literature suggests that research samples should also be led by farm size [49]. The average farm size and type in Northern Ireland is small/very small and livestock based [50], and the sample reflected this.

As illustrated by Errington and Gasson, you cannot fully understand the family farm structure without understanding the family relationships that manage it [8]. The roles of family farm members, and the relationships between them, can also heavily influence and/or aid family farm decision making [51]. Yet, women's roles in the decision-making process are often underestimated, despite their significance [52,53]. Chiswell also argues that while it is essential to understand the intergenerational process, it is also imperative to recognise the successor and who they are, as valuable actors in the retirement and succession process [54]. In some cases, there are family members who also want to farm but cannot, given that their sibling is first choice (generally because they are older). In this case, the '(non-)successors' are also important research participants, as their role is often undervalued within family farm continuity processes [55]. To respond to the significance of these different family roles, the fieldwork targeted the recruitment of farmers, farmers' spouses, successors, and (non)-successors for the in-depth interviews.

The final purposive sample was 62 participants, with 21 farmers aged 33+, 20 women aged 21+ married to farmers, 20 successors and one (non-)successor aged 18+, from 20 small/very small farms with livestock across NI. One-quarter of the participants interviewed were from the same families (e.g., a farmer, their spouse, their successor (adult) child). The other participants were independent family farm interviewees. All but two farmers were men, all spouses were women, all but two successors were men, and the (non-)successor was a woman. Out of the 62 farm participants interviewed, 53 were educated to technical or third-level college, with the remaining nine educated to the primary or secondary level (each of which were aged over 60 years).

The interviewees are assigned labels as illustrated in Table 1. In the discussion of the interviews, after each label (e.g., Farmer, Spouse), a number is included which reflects the order in which the participants were interviewed (i.e., Famer 1, Spouse 1, etc.). Given the dominance of men in the sample of farmers and successors, gender labels are only provided if the farmer or successor was a woman.

Table 1. Farming Participants by Cohort.

Farmer, aged 18–49	Spouse, aged 18–49	Successor, aged 18–29
Farmer, aged 50+	Spouse, aged 50+	Successor, aged 30+ (Non-)Successor, aged 18–29

Adopting a Grounded Theory approach [45,47], this research began with a sample of farming participants (farmer/spouse/successor/(non-)successor) who were identified and selected by the lead researcher by attending various farming events across NI. These events, and the interviewees who were recruited from them, allowed for a snowballing process to further inform, and in turn recruit, additional farming participants. While this simple referral system can have some disadvantages such as selection bias, these were outweighed by the opportunity to reach potential participants who would otherwise be difficult to establish contact with [56]. The initial sample of interviews allowed for the identification of emerging themes through an iterative process of interview data analysis until 'data saturation' occurred [45,46]. The priority of this research was to give farming participants a voice, in order to articulate and, in turn, better understand the emotional perceptions of decision making within changing rural communities. The methodology underwent ethical approval and interviews which were conducted from October 2019 to March 2020, through two phases.

3.2. Fieldwork Methodologies

3.2.1. 'Work and Talk' Interviews

The first research phase involved 21 oral life history 'Work and Talk' semi-structured interviews, which were conducted from late October 2019 until March 2020. The interviews were used to explore the participants' biographies, where interviewees could recall and reflect on their experiences in their own words, aiming to "move away from well-rehearsed, amusing anecdotes to a deeper exploration of subjectivity" [57] (p. 5). It has been suggested that this articulation of personal 'life stories' can prove a useful technique to encourage participants, and in particular older men, to open-up about their individual personal experiences [58,59].

The innovative 'Walking and Talking' methodology of Anderson (2004) is a popular tool in qualitative rural research [60] (as examples, see Riley) [61–64]. A fusion between interviews and 'hands-on' observation, this approach has grown in popularity for exploring themes related to the relationships between self, space and attachment, and usually takes place in an area relevant/related to the research [65]. As an example, Riley's study on changing agricultural practices adopted a Walk and Talk approach, and provided the interviewer with access to 'hidden voices' on the farm, as well as allowing for a flexible approach in order to fit into the everyday commitments of the respondents [66]. The embodiment of walking also had the added benefit of centring 'place' in the research; as Riley, notes, "The farm is a site of knowledge construction, and understandings may be embedded within, and layered on its fields and practices" [66] (p. 662).

Inspired by the benefits of this Walk and Talk method, a novel participatory approach was developed for this research: 'Work and Talk'. The lead researcher became immersed in the everyday life of the farm by working whilst interviewing. This involved, for example, shadowing and assisting each participant either on the farm or the family home by doing menial tasks such as brushing the yard, mucking out, or helping in the kitchen.

The Work and Talk approach proved productive in several key ways. As with the Walk and Talk method, it provided a flexibility with the interview process that allowed the interviewer to respond to the participants' working commitments on site. Farmers felt more able to agree to be interviewed, since the process was less distracting from their working day. The approach also unveiled 'hidden voices', where participants were free from judgement (see Riley [66]), as illustrated by Farmer 6 (aged 50+), when discussing issues around farm aspirations and retirement which he was reluctant to discuss as a couple earlier in the interview: "*I can say that now ... sure she would say the opposite!* [laughter;

indicating to the farmer's spouse in the distance]" ... As the lead researcher (who also conducted the interviews) is not from a farming background, there was initial concern over potential problems in gaining access to family farm participants, given negative perceptions of their 'outsider' status. As an example, Farmer 11 (aged 50+) remarked, while indicating to the researcher's attire, *"well you're not from farming anyway"*. Kuehne argued that it is important, when interviewing farmers, that the interviewer should try to 'fit in', and demonstrate their interest in the farming way of life, to encourage a good rapport [67]. Offering to conduct the interview while helping with basic farm tasks through this 'Work and Talk' process immediately softened any distrust or indifference towards the interviewer. The approach enabled the interviewer to earn respect by 'getting stuck in'. This not only gained the researcher access past the elusive farm gate, but also helped with further research sampling.

The embodiment of Working and Talking on the farm also gave insight to the relationships between the participants' life stages, across time but also *place*. It encouraged participants to open-up and reveal more of their biographies, exploring and reflecting on the physical family farm. Younger successors also engaged with this remembering and reflective process, where working and talking gave them a chance to 'show off' the farm, often illuminating a great sense of pride, especially if the farm had undergone improvements, such as the introduction of new milking equipment. This biographical participatory method, being undertaken physically on the farm, was also often viewed as a welcome excuse to stop and reminisce, especially for older farmers and their spouses. This was even more acute at the kitchen table, which also gave the interviewer insight into relationships and in particular power dynamics within the family, especially between the farmer and their spouse. It was through the natural flow of finishing work and walking to the home, within their safe space and surrounded by generations of family memorabilia, that the participants were more animated and emotional, encouraging a deeper exploration of the research themes. Some farmers would emotionally and wistfully recall their mother baking bread, or making stews for all who entered the house, especially when farm advisors were regular visitors. There were also happy, sad, or difficult emotional recollections of the farm while moving between sites, as discussed, for example, by Farmer 4 (aged 50+) *"I have no one to take over;* [crying] *my daughter has no interest in the farm and what do I do, I don't want to leave, look* [indicating to a photo of the generations of family on the farm]."

Each interview lasted one to three hours, and was audio recorded with permission and transcribed soon after the interview took place. Themes of belonging and rural community change are at the roots of this research, and questions were loosely asked around participants' perceptions of embodiment (space, place, attachment) to their farm decision making across their lifetime. Farming participants were not asked if they felt like they 'belong' to rural communities, but rather what 'belonging' on the farm and rural communities has meant to them, in their occupation and identity as farmers, and how belonging within changing rural communities may have affected their retirement and succession decision making. For example, participants were asked about their farming identity, and if and how it had changed within their lifetime. They were asked to reflect on the emotional aspects of family farm retirement, as well as questions related to family farming and community. A sample of the interview question guide can be found in Appendix A.

The use of the oral life history Work and Talk method greatly suited the emotional focus of this research, especially with the older farmer and women participants, as it helped guide them through any emotional past experiences that connected to the present. For example, one farmer was visibly upset in the interview when recalling how he had been asked by his spouse to knock down some farm sheds (due to aesthetic reasons) near the family home. Due to the generations of family and community 'ceilidh' dances that had been held there, the farmer was reluctant to destroy the buildings, and it was causing much conflict within the family. In this case, as with others, the interconnectedness between the farmer's physical and emotional attachment to the farm was revealed.

3.2.2. Online and Telephone Oral Life History Interviews

In March 2020, the initial first phase of the Work and Talk research was adapted due to COVID-19 limitations to protect the interviewer and research participants [68]. The second phase of the research fieldwork thus continued with the remaining 41 farming participants using oral life history online/telephone interviews, which took place from July 2020 until December 2020. Thirty of these remote interviews were conducted on the telephone using chat software with the remaining farmers and spouses. One farming couple took part in the telephone interview together, while the rest were individual. Reflecting differences in technological preferences/confidence between generations, the rest of the remote interviews were conducted with successors aged 18–29 using video conference technology.

Despite the advantages of the Work and Talk approach, there were some limitations, such as family members interrupting interviews, or distractions given participants' workloads on the farm. Additionally, unless successors had taken over the farm completely, or were heavily involved, the 'Work and Talk' approach was less appropriate for this group. Moving from face-to-face to a combination of online and telephone interviews helped with this issue; successors generally preferred the freedom to talk on their smartphones. This approach also provided successors with the opportunity to share and reflect on photos [69] of their farm, ensuring that place remained central to the conversations, despite the interviewer no longer being on site. As an example, Successor 7 (aged 30+) had inherited the farm from her uncle, and to facilitate a discussion of her farming passion through her life course, she shared images of childhood memories on the farm. The second phase also proved especially effective when discussing sensitive topics; telephone interviews can, for example, reduce stress in emotionally charged situations, allowing more freedom to disclose information because of the lack of visual representation [70]. As an example, Spouse 8 (aged 50+) admitted at the end of the interview that she had found it easier to talk to a stranger on the phone than she might have in person about their family conflict following her son's marriage breakdown. While this loss of visual interaction can be disadvantageous [70], a mixture of telephone and online interviews proved effective for reaching a diversity of participants. Remote interviews also allowed for flexibility in terms of when the interviews could take place, and their duration [71]. Some spouses preferred to take part in the interview late in the evening when they 'had time to relax', or their formal workload was complete. However, and importantly, participants now also had a *choice* of where the interview took place, enabling greater privacy, without other family farm members nearby. This tended to aid a more open conversation about the emotional aspects family farm relationships affecting family farm decision making.

3.3. *Analysis*

All semi-structured interviews were transcribed verbatim after each interview, and each transcript was then open line coded manually, by exploring recurring themes across the transcripts. Transcripts were then re-read, and axial coded through an iterative process, identifying relationships between the categories until theoretical saturation was reached [45]. These themes were then explored, informed by the work of Ajzen, Allen and Kern as well as that of Antonisch, Fenster and Yuvas-Duval, and Debeauvoir, as illustrated in Table 2 [10,30,72–74].

The transcripts were then also analysed by cohort of farmer, spouse, successor and (non-)successor, to compare emotional perceptions of belonging and decision making between the groups. Memo-writing, undertaken throughout the interview process in addition to the recording, helped to clarify any connections between codes and categories based on participants' body language, emotions, feelings, demeanour, etc.

Table 2. Research Conceptualisation: Emotions and Family Farming over the Life Course.

Unresolved or unrecognised emotions	To signal any habits or policies that require attention through changing rural communities or a changing family farm structure and have been excluded to date.
Farmers' Planned Behaviour	The traditional patriarchal structure, what is considered 'normal' by farmers and peers, their attitudes, intentions, beliefs and/or how easy it is to implement or take control over their decision making.
Emotional Sense of Belonging	Participants' emotions towards social bonds and the desire to connect, attachment to groups or people (including through childhood experiences), perceived quality of those social bonds.
Emotional Place Belonging	A sense of security that is built through a feeling of 'home' through participants' oral life history, social connections, cultural, economic and environmental factors.

The fieldwork was conducted in the uncertain times of Brexit and the global pandemic; however, the emotional focus of this research was notably welcomed by the interviewees. Once the farm gate was (literally and metaphorically) opened, most of the participants were eager to share their views and emotional experiences of family farming, often reflecting on the process as cathartic during what were difficult times for farmers and their families. There were even expressions of gratitude that this research was being undertaken, as illustrated by Farmer 1 and Spouse 1, who remarked *"it's good you are asking, this is great, this should've been done more before"*; *"you think no one cares"*.

4. Results

The following sections explore the core themes which emerged from the in-depth interviews: the changing patrilineal tradition identified through a 'longing for belonging' by keeping the farm in the family, communication on the family farm in relation to decision-making processes, and the emotional impacts of developing policies and relationships with policy makers.

4.1. A 'Longing for Belonging' to Keep It in the Family

As noted earlier, Northern Ireland has a unique relationship with issues of land ownership, and this emerged very clearly from the interviews. A passion for the land and the farming 'place'—the family farm and its surrounding rural landscape—emerged clearly from the farmers' narratives. This had a particular resonance with farmers for whom the land had been passed down through familial generations. There was an awareness that farmers felt a sense of security, or a 'feeling of home', through 'place belongingness' [14], by keeping their farm in the family.

Farmer 1, who has two daughters and one son, provides an example of this emotional security. For him, as with other participants, keeping the farm in the family was very important, but his son was not interested in the farm. His spouse, however, was adamant that each of their children would be given an equal share of the farm. Regardless, he still holds on to hope for this strong traditional patrilineal to be maintained, and that his son would eventually change his mind and take over the farm completely, regardless of any conflict with his spouse or daughters:

"It would be a very simple decision if [son] had the interest in the farm and nothing or not really nothing else, but he had a real genuine interest in the farm. I would not be intent of leaving [son] a third of the farm, or half of the farm, or two thirds of the farm, ... he gets the lot" (Farmer 1, aged 50+).

Farming participants' 'sense of belonging' and rootedness to place was evident through the pride shown across generations and the desire to retain the farm in the family.

However, in some cases, familial conflicts had arisen whereby (potential) successors were less interested in inheriting the farm. Two successors, for example, described the tradition of keeping the family name on the land a 'hindrance', and, while expressing pride in their family's land, reflected on how they were not 'sentimental' about it.

Yet, a passion for family farm tradition was also evident for several of the successors, who felt great comfort from the generations of family ownership on the land. For example, two successors were afraid of losing their farmland, not just physically, but also emotionally. In particular, their emotional relationships with their farmland related to themes of memory, family, inheritance, knowledge transfer, and belonging:

> *"Is it true, is it really true that I will not get to live here? Because if that is the truth, then you know I may as well just reconcile and let this five-generation thing go ... but it's very, very hard, because even as I sit here, I have memories of being a child, I get great comfort being here and it has helped me with his [uncle] loss"* (Successor 8, aged 30+).

> *"I think it's key to any farm that the family is involved in it, you know, ... anybody can learn how to farm from nature ... but it's from your family you get most stuff"* (Successor 19, aged 18–29).

The emotional connections to land, and to farming as a way of life, were expressed throughout the interviews with farmers. In reflecting on his own experience of finding an alternative occupation, one farmer argued that if farming is 'in your blood', it will not go away:

> *"Well, my parents did everything they could to stop me farming, I was working with an accountancy firm for four years, and then I realised it was the love of farming that brough me back"* (Farmer 17, aged 50+).

Yet, despite the desire to maintain the farming tradition, and farm space, in the family, farmers and their spouses commonly explored in the interviews a conflicting push–pull dynamic. The pull of strong emotional attachments to the farm were accompanied by a tension in not wanting to place this traditional burden on their children and the subsequent generations, with some families viewing the farm instead as a 'poisoned chalice':

> *"I think in some areas there's still a big attraction to keep it in the family, or a responsibility to keep it in the family in certain areas, but I think that's all changing now"* (Farmer 7, aged 50+).

> *"To be completely honest I think it [keeping the family farm] interferes in the family, but my husband would like it"* (Spouse 12, aged 50+).

> *"It's nice, but as I said, I have said to [son] quite openly, if this place isn't working for you and you saw a nice block of land ten, fifty or hundred miles away and you know, move; don't let this history hold you"* (Farmer 2, aged 50+).

A complicated relationship between the younger successor and the older traditional farmer also illuminated contrasting views towards the tradition of keeping the name on the land. Successors, when asked about this tradition and their plans for the future of their family farm, used language to describe older farmers as 'set in their ways', and that it is 'their way, or no way'. Farmers' view of the successors to the same question, on the other hand, was of successors having no work ethic or passion for the land. These seemingly strained relationships were causing conflict for some older farmer participants, and illustrated a lost 'sense of belonging' through a lack of confidence in the next generation and being able to pass the farm through traditional intergenerational transfers, as Farmer 4 (aged 50+) described:

> *"You know, the general hearsay around the country today [is] that farms generally only last three generations in any one family ... the first generation buy the farm and they live in debt for the rest of their lives. The next member of the family takes it over. The next generation—he finishes up any payments and he develops the farm up to what you*

know to be a successful business, and then the third generation comes along and, excuse the phrase, he pisses it up against the wall! [laughs]."

Indeed, some successors were open about not being as passionate about the farm as their predecessors, or keen to work the same long hours. The pull of the emotional attachment of place belonging associated with keeping the name on the land was instead more pragmatic for many successors, where they felt that the accomplishments and sustainability of the family farm today were based on economic choices and not on traditions or legacy [75]. This is well illustrated by the following comment from Successor 7 (aged 18–29):

> *" . . . the family name on the farm is very important . . . but if it came down the line and you had to move or something . . . I would like to keep it in the family name, but, you know circumstances with business decisions—you have to make a move . . . it wouldn't be the end of the world sort of thing."*

However, the patrilineal pressure to keep the name on the land was shared by those successors who had a clear emotional attachment to the farm. In some cases, there was evidence of experiences of stress in keeping the farm going through the family bloodline. Some successors expressed how they felt an incredible responsibility to prevent this loss of family tradition:

> *"You just wouldn't like to think that the whole thing ends with you—that would be my main fear"* (Successor 10, aged 18–29).

> *"Well, I suppose when you have such a long line of it going back, you sort of feel a bit of responsibility. [Interviewer: is that a lot of pressure?] . . . a lot of pressure, but you don't really think about it too much, but when you do there is pressure"* (Successor 10, aged 18–29).

The deep interconnections between emotional connectedness to farming and the land, and the emotional pressures of maintaining the family farm, were reflected on in an interview with one successor who tried hard to keep the farm in the family through five generations. Successor 8 had worked on the farm with her uncle for ten years and had nursed him before he died. However, she explained how she quickly felt pressure, intimidation and aggression from her neighbouring uncle and his sons to leave the family farm, given her position as a woman, and a niece, and thus as someone deemed an inappropriate inheritor:

> *"He [uncle] was in hospital a few weeks later. He talked to me about this [potential problems with family members regarding inheritance] in depth and he said to me, and his last words were, 'I don't think you're going to have peace to live there', and he took a heart attack . . . [sobbing] and I lost him to that . . . and if that wasn't enough, the way we buried him and it was like you know, I feel like I was a pariah in the graveyard and I was really low on the wall and the family were all the way up on the hill looking down [silence]."*

> And

> *"In October I was out on the road with a measuring tape getting ready to measure fencing and my [other] uncle swerved the car right at me on the road right for me . . . I am not kidding you"* (Successor 8, female, aged 18–29).

The desires, and pressures, to hold onto the traditional patrilineal line on this 20 acre farm ultimately resulted in much family distress and an ongoing court case. There was evidence with this successor of a loss of emotional security, built by generations on the farm, because they did not conform to the traditional patrilineal family farm inheritance structure.

Along with changing views towards traditional farming, a shift in relation to family farm communication and decision making also emerged from the narratives, as is next explored.

4.2. Family, Communication and Farm Decision Making

One of the core ways in which the emotional aspects of farming retirement and succession decision making were revealed was in the discussions around family communication. Most participants described family conversations in very positive terms, reflecting on how all farm issues were discussed together either at the kitchen table, or on the farm while working. This was discussed by Successor 6 (Female, aged 18–29):

> *"It's quite simple at home. We all work together: me, Daddy and my sister ... We make decisions together. Daddy involves us heavily in the farm as well. Like, he would ask our opinions ... we all get to make decisions and things like that ... And we all work hard on the farm today. So, at the moment, it's just all three of us pulling together."*

It was also agreed by most participants that while the determinants of farming succession had multiple economic dimensions, the emotional aspects of this decision making were also hugely important. These involved discussions with the interviewer around issues such as poor health, anxieties where there was no successor in line, and, more generally, the emotions of retirement planning. Only one-quarter of the participants had any retirement plans or wills put in place. It was evident that this remained a very emotive issue to discuss amongst family farm members (see also UFUNI [6]). Even when there was a will in place, the details of retirement plans were not always shared with, or communicated to, the successor:

> *"[Interviewer:] Have your Dad and you sat down and talked about succession or retirement plans?*
>
> *[Participant:] Well, no not directly like ... he has a will and things created and sorted if things go wrong, but there's not an actual sort of time-line as to what will take place like"* (Successor 4, aged 30+).

Some older farmers noted the importance of having a will in place, but they felt there was not enough government support or advice to encourage and aid farmers through this emotional process.

The narratives revealed how the most dominant influence on family farm decision making was family farm relationships; an essential element in understanding family farming [76]. This was often strongly linked to themes around perceived threats to the farmers', and in some cases, their spouse's, 'sense of belonging', especially if they were not from a farming background [35]. A prominent example that emerged in these farm relationship discussions, across the three cohorts of farmers, spouses, and successors was the changing views around martial traditions and dissolution. Many of the older interviewees explored issues of stress, anxiety, and fear around the potential for, or in some cases, realisation of marriage breakdowns of successors and how it would impact the future of the farm.

> *"There's a situation ongoing up the road there; the wife has left the husband after only about three or four years of marriage, and we don't know what way the whole place is going to end up"* (Farmer 20, aged 50+).

> *"My nephew had a marriage that broke down and it was a really dirty breakup ... his father who was a farmer would say that they were very lucky not to have signed over anything or it would all be gone"* (Spouse 8, aged 50+).

One farmer and his spouse had experience of the loss of the farm after their son and daughter-in-law separated. They viewed their daughter-in-law as an 'outsider' given her lack of farming background, and consequently placed the blame on her over the marriage breakup and poor communication from their son about the subsequent selling-off of part of the farm:

> *"[crying] it has broken the family ... none of my daughters talk to him anymore. We don't know what to do now—we thought we could retire. I worry about it all the time"* (Spouse and Farmer 7, 8 (joint interview), aged 50+).

One farmer (aged 50+) coined the phrase the 'dreaded daughter-in-law' to describe these family dynamics, the discussions around which were often highly gendered, and strongly linked to themes of entitlement and (especially when the women were not from a farming background) perceived threats to their place belonging and identity (see also Price and Evans [35]). Several farmers and their spouses reflected on the emotional stress of retirement decision making in the context of these relationships:

> "*A lot of farmers don't want to hand over their property to their son because of a particular reason. Down the years when my father was living and when I got married marriage was for life, ye know what I am saying! A lot of the things are going belly up at the minute and a lot of farmers would prefer to keep the next generation in line on the farm*" (Farmer 10, aged 50+).

> "*This is not a judgement comment, but as more and more farmers marry non-farming daughters, that has a big impact as well. Because those ladies don't understand this 80-hour week, week after week, and they shouldn't ... but, you know, the in-law thing ... if my son said 'I want to work Monday to Friday, you know 50 hours a week, will you do weekends?' I'd think, no I bloody won't!*" (Farmer 2, aged 50+).

> "*My biggest worry is my brother getting married, and, you know, I want to see it, but I don't really have that relationship with his future wife. And I am kind of worried that, you know, you hear of all these farms being sold because of divorces and everything, you know. That's kind of my biggest fear at the minute ... If my brother marries, I am dreading it; my sister-in-law is a driving force*" (Spouse 18, aged 18–49 (farm is split three ways with brothers)).

Successor 2 (aged 18–29) believed that marriage breakdowns were also related to suicides in farming:

> "*If they split up in a marriage that's when ... that's when suicides and things come up.*"

This fear, and lack of trust, was relayed powerfully by one successor, who explained how, at the age of 44, he did not want the farm signed over to him in case of a future marriage breakdown and potential loss of the farm:

> "*I really am in no rush for them to sign over the whole farm because there have just been so many incidences with farmers in around my age where the wife has up and left, and it's been a lot of hassle ... so I am not given them any pressure on my mother and father to do it, 'cause in two years' time I could be happily married but you never know.*"

The threat of marriage breakdown was magnified because of the reported lack of legislation in place to protect generations of land ownership. Strong emotional responses to discussions around this issue related to themes of unfairness and frustration, particularly in relation to intergenerational land ownership:

> "*I think it's wrong, that a wife is able to claim half of the farm ... I could see it happening here ... But anyway, I think that's awful, I think it's dreadful that a man who has built up a farm all his life, his father before has built up the farm, he has handed it over to his son, something has happened, and the wife walks off with half of it. It's not right!*" (Spouse 9, aged 50+).

While much of the discussions on communication and relationships focused, unsurprisingly, around the farming family, some farming participants also explored the relationship and role that the animals had on the farm had in yielding a strong sense of belonging. There was an evident connection between the attachment of farm animals to the farm and family [77,78] and 'linked lives', especially with the older farmer [79]. As one farmer reflected:

> "*[Interviewer:] Are you attached to your livestock?*
>
> *[Participant:] [cried] I sold two cows recently, and it knocked me for six that they were going to be culled ... they were my friends; I work with my friends I didn't want to sell my friends ... it makes me really sad*" (Farmer 7, aged 50+).

The emotional pain felt by older farmer participants when this relationship is lost also affected family farm decision making (see Tovey, 2002) [77]:

> *"My cows, my children, they all have names ... ahh, yep, the fact the contract fella bought the herd, and they are staying here 'cause I rented the old buildings too ... I didn't have to go through the pain of getting rid of cows; we lost 40 cows almost six years ago to TB, and it was like a death in the family for me and my wife"* (Farmer 9 aged 50+).

However, the links between emotional attachment to the farm and animals were not held by all, and some expressed indifference to any attachment to animals and its implications for retirement decision making:

> *"No attachment at all. Sure, you can't—you're not a retirement home for cows you know! You're a business!"* (Farmer 10, aged 50+).

The results indicate that it is essential to consider farmers' relationships with their animals in the farm decision-making processes [78], including their emotional bonds, security and attachment, which are often reinforced daily on the farm. In turn, this showing of compassion for their animals, and being viewed as the 'good farmer', has an important role in shaping farmers' identity [79].

4.3. *The Emotional Impacts of Policy*

Across the three cohorts, there were continued aspirations for the family farm. As examples, some successors planned to enlarge the farm, and to introduce new technology such as solar panels and robotics milking; farmers' spouses often encouraged diversification, and, if from a farming background, managed the economics of the farm; some farmers were hoping to rent out the farm, or contract it to a younger farmer, enabling them to retain the land, but also to retire. Yet, many of these plans were discussed with a heavy heart, with strong feelings that there is little help to achieve them. As Farmer 16 (age 50+) explained:

> *"I really want to find a young farmer to take over the farm as my girls don't want it. Someone so I don't have to sell on the herd, and they can just take over, but I can't, and there is no help—I am really stuck."*

The interviews involved emotional discussions around feelings of being unsupported. The Department of Agricultural, Environment, and Rural Affairs (DAERA, 2021) is the government body in NI tasked with overseeing food, farming, environmental, fisheries, forestry, and sustainable policies. It also acts as the local managing authority for delivering and evaluating all rural development programs to support growth, jobs, and sustainable development in rural areas [80]. The Department was a core focus of many of the discussions around the perceived lack of farming support, and the emotional burden of feeling unprotected.

> *"I think a sense of hope, you know, with policy the way that it is at the present time. Yeah, there's a policy of no hope ... people feel very vulnerable at the present time; we are being told so many mixed messages"* (Farmer 4, aged 18–49).

> *"We have a department, and we have people now who are making decisions, who are looking at a screen and they have a little block graph and things as to what's best for people. They don't consider the emotional side of things—how things should be done. They just say what's possible legally, and we can do that because as an EU Directive we have to implement this particular law or system: you know, like it or lump it"* (Farmer 4, aged 18–49).

A theme of disconnect between farm families and policy makers was evident across the three cohorts, where participants were often animated, upset and angry in interviews, expressing how they felt little respect from the body that had helped them so effectively in the past. Interviewees argued that this had a direct impact on family farm decision making. Poor communication was a recurrent theme, as well as frustration by older farmers and their partners who felt unable to cope with changing processes:

" ... DAERA now want you to do everything on the computer online ... before, the farm advisor would come out and sit and talk to you. There is no personal contact at all" (Spouse, aged 50+).

"When I started being a farmer in 1982, I can still remember a great Department of Agriculture personnel who helped my farm develop, and if there was a grant application, a bit of guidance to help me, lots of things, they really helped. That person does not exist anymore within the Department" (Farmer 15, aged 50+).

"DAERA needs to go back to its roots and have a more emotional connection with the family farm like it did in the 1980's" (Farmer 9, aged 50+).

The consensus amongst interviewees was that these changes were for the negative, with one successor describing representatives in the government body as 'pencil pushers in their ivory tower' (Successor 13, aged 18–29), who were no longer in touch with the farmer and what the family farm needed. Despite agricultural, and more broadly rural, developments over the last few decades, farming in NI remains fairly traditional, and many of the older farmers felt that they needed personal, and preferably physical, communication with advisors:

"Farming is such a different occupation, and a different business, and it's something you have to visually see and understand, and then you have to be from it as well, I think ... 'You would [in the past] have an advisor and they would walk the farm with you, and they had come up with ideas and seeing things through their eyes. But now you just have to apply it all online and there's something seriously missing there" (Farmer 3, aged 50+).

The interviewees' narratives explored how they felt that the government had lost touch with farming families, and as a result has lost respect towards the family farm, despite more positive relationships in the past. These discussions reflected on themes of loss, particularly around their connectedness to the Department, and the emotional security that it had once provided. The farmers and successors interviewed admitted to becoming unengaged with policy advice, which was potentially seriously damaging for farm decision-making support. As Farmer 6 (aged 50+) noted:

"I've been at a few of these seminars, these retirement successions, and these farmers come along and they're lovely—the best of the world—and they have fear, and you could see them 'cause I'm looking from the outside looking in, and you can see they're scared, but when they leave the room, they don't take anything with them."

5. Conclusions and Implications

This research set out to pay attention to the emotional aspects of farming decision making around retirement and succession. Drawing on a series of in-depth oral life history interviews with farmers, their spouses, and successors, this article explored how family farm participants' emotions towards place, and their sense of place belonging, influences family farm decision-making processes [9].

The interviews revealed how farmers, and their families, depend on their surroundings and farm relationships (explored broadly, including family, animals, place, and policy makers) to understand their emotional and physical identities, but felt that these identities were being threatened. The farmers had a strong sense of place belonging in their farmland, because it is here where most participants spent most of their time, where they physically worked their land, and where social and familial bonds were made and strengthened. It also became clear through the participants' narratives that there was enduring emotional security and comfort gained from generations of family on the land, where a consequent 'pride of place' was forged. However, this passion and reliance on the traditional family farm were often subject to a push–pull dynamic. Alongside this pride, there was a changing tide towards more pragmatic views of responsibilities towards retaining the family name on the farm. This was evident across all cohorts (farmers, spouses, and successors); the interviews revealed anxieties around not wanting to pass on a 'poisoned chalice', and explored generational work ethic differences. Yet, these changing attitudes were not always shared, and there were cases where a 'longing for belonging', and its relationship with the survival of the farm, and farming name, created conflict between farmers and successors. Despite significant agricultural or rural developments, attachment to the land and the desire to keep the farm in the family, remained persistent. This longing for belonging was multipronged and complex, and the emotional aspects of these preferences, and often pressures, revealed the intergenerational and gendered relations involved in retirement decision making. From a policy perspective, the discussions with farmers and their families strongly suggested that a greater understanding of the emotional aspects of farming decision making could act to rebuild relationships between the government department responsible for rural affairs and farming communities. Our work with farming families suggested that further research is needed on the links between the emotional impacts of decision making in family farming and the health and well-being of farming families, particularly in order to encourage rural sustainability.

An innovative 'Work and Talk' oral life history methodology provided greater insights into the important relationship dynamics on the family farm, through first-hand observation of family farm roles. This participatory approach highlighted, through the embodiment of working and talking (see Riley in relation to walking and talking [61–64], the methodological advantages of encouraging participants to recollect by linking place—their farm—through time and their emotional experiences. This research has advocated for a greater appreciation of the emotional aspects of family farming, which are often overlooked at the expense of economic factors. By explicitly paying attention to the emotions of farming families, the interviews revealed the emotional social activities and individual emotional securities built through a sense of place belonging and how these can benefit (e.g., through enhanced support) or challenge (e.g., through additional pressures) family farm decision making. The findings resonate with Conway et al.'s claim that it is the 'soft' issues in family farming that are the 'hard' issues affecting retirement and succession decision making [81]. By focusing specifically on the relationships between place belonging and the challenges faced by farmers in deciding to 'let go' of the family farm, and/or to retire, our analysis contributes to an emerging body of rural research which aims to explore emotions across a diversity of issues impacting family farm decision-making processes.

Author Contributions: Conceptualisation, L.A.H., G.C. and A.S.; methodology, L.A.H., G.C., R.N. and A.S.; formal analysis, L.A.H.; writing, review and editing, L.A.H., G.C., R.N. and A.S.; Supervision, G.C., R.N. and A.S. All authors contributed to writing, editing and reviewing the draft submitted. All authors have read and agreed to the published version of the manuscript.

Funding: This research was funded by the Department of Agriculture, Environment and Rural Affairs (DAERA) as a Ph.D. scholarship awarded to Lorraine Holloway in September 2016, mail to: karen.patterson@daera-ni.gov.uk.

Institutional Review Board Statement: The study was conducted according to the guidelines of the Declaration of Helsinki, and approved by the Institutional Review Ethics Committee at Queen's University Belfast (EPS 19-250 updated on 1 June 2020).

Informed Consent Statement: Informed consent was obtained from all participants involved in this study.

Data Availability Statement: The data are not publicly available for confidentiality reasons and in line with ethical considerations.

Acknowledgments: The authors are grateful to the Department of Agricultural and Rural Affairs (DAERA) for funding and advice. Sincere thanks to the guest editor John McDonagh for preparing this Special Issue. Thank you to the anonymous peer reviewers, for their valuable feedback on previous drafts and Linda Price's generous support with the initial stages of this project. Lastly, we would like to pay our gratitude and respect to our dear co-author and colleague, Aileen Stockdale, who passed away suddenly in March 2021. Aileen was instrumental in the creation of this research project and championed it until the end with great enthusiasm and valuable direction. Aileen was not only an outstanding academic in her field, but she was also a dear friend to many who was always generous with her time (and great sense of humour). Aileen was constantly in our thoughts for this Special Issue and will be forever missed but not forgotten.

Conflicts of Interest: The authors declare no conflict of interest.

Appendix A. Sample Interview Question Guide

Farming Background:

How long have you been in your role as farmer/successor/spouse?
    Are you satisfied with your role?
Did you grow up on a farm?
    If yes, level of involvement on farm?
    If no, what were your previous experiences of farming?
How did you acquire the farm/role? (bought/inherit/marriage/succession)
    Were you involved in this decision?
If inherited, how many generations of your family has farmed this land?
Did you want the role?
    If no, what did you want to do?
    What influenced your decision to then farm?
Is farming your only job? (FT/PT/Hobby)
    If no, what is your other job?
    What influenced(s) this decision?
Do you think farming as a 'way of life' has changed over your life-course? How?
Do you think it is important to keep the farm in the family? (patrilineal) Why? Has this opinion changed?

Family Farm Decision Making:

How are farm decisions made and communicated on the family farm?
    Who do you think is the principal decision-maker on the farm?
    How involved are you in the farm decision-making process?
    Do you discuss farm issues as a family?
    Would you like more say in the decision-making process?
    Has the farm decision-making process changed over the years?

Succession Decision Making:

What have been the main influences on succession decision-making in your farm household to date?
    Are there any (other) external/internal influences?
Is there anything you would like to change about how succession decision-making is communicated/decided in your farm household?
Who do you think is the lead farmer—older farmer or successor?
    If not you, do you want to be? And what do you think the challenges are to accomplish tthis?
    What are your challenges and fears for the family farm?

Retirement Decision Making:

They say farmers don't retire—do you agree with this statement?
What have been the main influences on retirement decision making in your farm household to date?
    Have there been any (other) external/internal influences?
How has this changed over your life-course?
What is your role in the retirement decision-making process?
How are these retirement decisions communicated within the family?
As successor, do you have any say in the retirement decision-making process? e.g., when you take over the farm officially. If not, would you/how?
Do you think farm retirement decision-making is more emotional for farm families than any other occupation?

Farming Identity & Sense of Belonging in Place:

How do you see yourself as a 'farmer'/spouse? (i.e., caretaker of the land). Has this changed in your lifetime?
Do you think the farming identity has changed? (any change in roles, values?)
What do you think are the main values of a farmer? Has this changed?
How attached are you as farmer to the farm today? Has/how has this changed in your lifetime?
How attached are you to the livestock/nature on the farm? Does this attachment affect your decision to retire/succeed?
Is your home a form of attachment to your role as farmer/successor/spouse?
If you do stay in the family home, is it important to still have a role on the family farm?

If you can't stay in the family home, what will you do?
Do you think to remain living on the farm in retirement is a good way to ease the transition out of farming? Does it help with the stress of retiring?
A Sense of Belonging in Rural Communities:
Do you agree with the statement that the family farm is the heart of rural communities today? Why and has this changed in your life-course/how?
Do you feel part of your local community? Has this changed over the years/why?
Do you think the farm reflects who you are in your rural community, networks, relationships? Is this important and has this changed over your life-course?
Do you socialise in your community/how?
What other clubs/networks/unions ave you been part of, if any? How often did you meet and is this important to you? Has this changed over your life-course?
If you do not socialise, what do you do? Has this changed? How does this make you feel?
How do you feel about rural communities today?
Do you think rural communities have changed? Do you think this affects your decision to retire or stay on the farm?
Do you think traditional family farms have kept up with modern rural communities?
Is there anything you think that would help farmers integrate better into changing rural communities?
Retirement/Succession Policies:
How do current rural development policies help/encourage you as a farmer in the retirement/succession process? How has this changed in your life-course?
Are you aware of all current policies to support retirement/succession? How are they communicated to you (if any)? Your views on the success of these policies?
Have there been any policies in your lifetime that you think helped the retirement/succession process?
How do you think the government might help farmers through retirement/succession planning and encourage younger farmers back into the farming industry?
What support do you think is needed for a smoother retirement/succession on the farm?
Are there any other issues with rural development policies or the retirement and succession decision making process that you would like to discuss?

References

1. Lowder, S.K.; Sánchez, M.V.; Bertini, R. *Farms, Family Farms, Farmland Distribution and Farm Labour: What Do We Know Today?* Food and Agriculture Organization of the United Nations FAO: Rome, Italy. Available online: https://www.fao.org/3/ca7036en/ca7036en.pdf (accessed on 2 November 2021).
2. Agricultural Survey. Available online: https://www.daera-ni.gov.uk/sites/default/files/publications/daera/Results%20of%20the%20December%202018%20Agricultural%20Survey.pdf (accessed on 20 June 2020).
3. *Farming in Northern Ireland*; Ulster Farmers Union: Belfast, UK. Available online: https://www.ufuni.org/farming (accessed on 2 November 2021).
4. Calus, M.; Huylenbroeck, G. The persistence of family farming: A review of explanatory socio-economic and historical factors. *J. Comp. Fam. Stud.* **2010**, *41*, 639–660. [CrossRef]
5. EU Farm Structure Survey. Available online: https://www.daera-ni.gov.uk/articles/eu-farm-structure-survey (accessed on 3 November 2021).
6. *Succession Planning in Northern Ireland Report, A Survey of Farmers Plans and Intentions.*; Ulster Farmers Union: Belfast, UK; Available online: https://content17.green17creative.com/media/99/files/SUCCESSION-ARTWORK-DOUBLE-PAGES_webfinal_2.pdf (accessed on 1 November 2017).
7. Bryant., L.; Garnham., G. The fallen hero: Masculinity, shame and farmer suicide in Australia. *Gend. Place Cult.* **2015**, *22*, 67–82. Available online: https://www.tandfonline.com/doi/abs/10.1080/0966369X.2013.855628 (accessed on 2 March 2018). [CrossRef]
8. Errington, A.; Gasson, R. Labour use in the farm family business. *Sociol. Rural.* **1994**, *34*, 293–307. [CrossRef]
9. Rieple, A.; Snijders, S. The role of emotions in the choice to adopt, or resist, innovations by Irish dairy farmers. *J. Bus. Res.* **2018**, *85*, 23–31. [CrossRef]
10. Antonsich, M. Searching for belonging—An analytical framework. *Geogr. Compass* **2010**, *4*, 644–659. [CrossRef]
11. Wood, N.; Smith, S.J. Instrumental routes to emotional geographies. *Soc. Cult. Geogr.* **2004**, *5*, 533–548. [CrossRef]
12. Simonsen, K. Practice, Spatiality and Embodied Emotions: An Outline of a Geography of Practice. *Hum. Aff.* **2007**, *17*, 168–181. [CrossRef]
13. Van der Veen, M. Agricultural innovation: Invention and asoption or change and adaption? *Word Archaeol.* **2010**, *42*, 1–12. [CrossRef]
14. Anderson, K.; Smith, S. Editorial: Emotional geographies. *Trans. Inst. Br. Geogr.* **2001**, *26*, 710. [CrossRef]
15. Pini, B.; Mayes, R.; McDonald, P. The emotional geography of a mine closure: A study of the Ravensthorpe nickel mine in Western Australia. *Soc. Cult. Geogr.* **2010**, *11*, 559–574. [CrossRef]

16. Ramirez-Ferroro, E. *Troubled Fields, Men, Emotions and the Crisis in American Farming*; Columbia University Press: New York, NY, USA, 2015.
17. Price, L.; Conn, R. *'Keeping the Name on the Land': Patrilineal Succession in Northern Irish Family Farming. Keeping It in the Family, International Perspectives on Succession & Retirement on Family Farms*; Lobely, M., Baker, J., Eds.; Ashgate Publishing Ltd.: Surrey, UK, 2012.
18. Conway, S.F.; Mcdonagh, J.; Farrell, M.; Kinsella, A. Cease agricultural activity forever? Underestimating the importance of symbolic capital. *J. Rural Stud.* **2016**, *44*, 164–176. [CrossRef]
19. Conway, S.F.; McDonagh, J.; Farrell, M.; Kinsella, A. Uncovering obstacles: The exercise of symbolic power in the complex arena of intergenerational family farm transfer. *J. Rural Stud.* **2017**, *54*, 60–75. [CrossRef]
20. Glover, F.; Reay, T. Sustaining the family business with minimal financial rewards: How do family farms continue? *Fam. Bus. Rev.* **2013**, *28*, 163–177. [CrossRef]
21. Grubbstrom, A.; Eriksson, C. Retired farmers and new land users: How relations to land and people influence farmers' land transfer decisions. *Sociol. Rural.* **2018**, *58*, 707–725. [CrossRef]
22. Shamai, S. Sense of place: An empirical measurement. *Geoforum* **1991**, *22*, 347–358. [CrossRef]
23. Buffel, T.; De Donder, L.; Philipson, C.; De Witte, N.; Dury, S.; Verte, D. Place attachment among older adults living in four communities in Flanders, Belgium. *Hous. Stud.* **2014**, *29*, 800–822. [CrossRef]
24. Longhurst, R. Introduction: Subjectivities, spaces and places. In *Handbook of Cultural Geography*; Anderson, K., Domosh, M., Pile, S., Thrift, N., Eds.; Sage: London, UK, 2003; pp. 283–289.
25. Stockdale, A.; Theunissen, N.; Haartsen, T. Staying in a state of flux: A life course perspective on the diverse staying processes of rural young adults. *Popul. Space Place* **2018**, *24*, e2139. [CrossRef]
26. Erickson, L.D.; Sanders, S.R.; Cope, M.R. Lifetime stayers in urban, rural, and highly rural communities in Montana. *Popul. Space Place* **2018**, *24*, e2133. [CrossRef]
27. Stockdale, A.; Ferguson, S. Planning to stay in the countryside: The insider-advantages of young adults from farm families. *J. Rural Stud.* **2020**, *78*, 364–371. [CrossRef]
28. Ryan, K. Incorporating emotional geography into climate change research: A case study in Londonderry, Vermont, USA. *Emot. Sp. Soc.* **2016**, *19*, 5–12. [CrossRef]
29. Wright, S. More-than-human, emergent belongings: A weak theory approach. *Prog. Hum. Geogr.* **2015**, *39*, 391–411. [CrossRef]
30. Allen, K.-A.; Kern, M.L. The need to belong. In *School Belongings in Adolescents*; Allen, K.-A., Kern, M.L., Eds.; Springer: Singapore, 2017; pp. 5–12.
31. Bryant, L.; Pini, B. *Gender and Rurality*; Routledge: New York, NY, USA; London, UK, 2011.
32. Allen, K.-A.; Kern, M.L.; Rozek, C.S.; McInerney, D.M.; Slavich, G.M. Belonging: A review of conceptual issues, an integrative framework, and directions for future research. *Aust. J. Psychol.* **2021**, *73*, 87–102. [CrossRef] [PubMed]
33. Hendry, L.B.; Mayer, P.; Kloep, M. Belonging or opposing? A grounded theory approach to young people's cultural identity in a majority/minority societal context. *Identit. Int. J. Theory Res.* **2007**, *7*, 181–204. [CrossRef]
34. Fenton, E.; Shields, C.; Mcginn, K.; Manley-Casimir, M. Exploring Emotional Experiences of Belonging. *Workplace* **2012**, *19*, 40–52.
35. Price, L.; Evans, N. From stress to distress: Conceptualizing the British family farming patriarchal way of life. *J. Rural Stud.* **2009**, *25*, 1–11. [CrossRef]
36. Riley, M. 'Letting them go'—Agricultural retirement and human–livestock relations. *Geoforum* **2011**, *42*, 16–27. [CrossRef]
37. Riley, M. Silent Meadows: The uncertain decline and conservation of hay meadows in the British landscape. *Landsc. Res.* **2005**, *30*, 437–458. [CrossRef]
38. Agricultural Census Northern Ireland. 2009. Available online: https://www.daera-ni.gov.uk/publications/agricultural-census-northern-ireland-2019 (accessed on 22 April 2020).
39. Agricultural Census Northern Ireland. 2020. Available online: https://www.daera-ni.gov.uk/news/final-results-june-agricultural-census-2020 (accessed on 1 November 2021).
40. Brexit: Future UK Agriculture Policy. 2018. Available online: https://www.parliament.uk/documents/commons-library/Brexit-UK-agriculture-policy-CBP-8218.pdf (accessed on 7 July 2020).
41. Brexit and Agriculture in Northern Ireland, House of Commons, Northern Ireland Affairs Committee, Fifth Report of Session 2017–2019, House of Commons, HC 939. Available online: https://publications.parliament.uk/pa/cm201719/cmselect/cmniaf/1847/184702.htm (accessed on 1 November 2021).
42. Price, L.; Simpson, M. The trouble with accessing the countryside in Northern Ireland. *Environ. Law Rev.* **2017**, *19*, 183–200. [CrossRef]
43. A History of Land Tenure Arrangements in Northern Ireland, University of Maine. Available online: http://dlc.dlib.indiana.edu/dlc/bitstream/handle/10535/4180/A_History.pdf?sequence=1 (accessed on 14 March 2020).
44. Feehan, J. *Farming in Ireland, History, Heritage and Environment*; University College Dublin: Dublin, Ireland, 2003.
45. Charmaz, K. Grounded Theory. In *Contemporary Field Research*; Emerson, R.M., Ed.; Waveland Press, Inc.: Long Grove, IL, USA, 2001; pp. 335–352.
46. Charmaz, K. *Constructing Grounded Theory: A Practical Guide through Qualitative Analysis*; Sage: Thousand Oaks, CA, USA, 2006.
47. Creswell, J.; Path, C. *Qualitative Inquiry and Research Design, Choosing Among Five Approaches*, 4th ed.; SAGE: London, UK, 2018.

48. December Agricultural Survey 2018. Available online: https://www.daera-ni.gov.uk/publications/december-agricultural-survey-2018 (accessed on 12 November 2017).
49. Burton, R.J. Seeing through the 'good farmer's' eyes: Towards developing an understanding of the social symbolic value of 'productivist' behaviour. *Sociol. Rural.* **2004**, *44*, 195–215. [CrossRef]
50. Statistical Review of Northern Ireland Agriculture 2019. Available online: https://www.daera-ni.gov.uk/sites/default/files/publications/daera/Stats%20Review%202019%20final.pdf. (accessed on 7 July 2021).
51. Registrar General Northern Ireland Annual Report 2016. Available online: https://www.nisra.gov.uk/sites/nisra.gov.uk/files/publications/RG2016.pdf (accessed on 27 March 2019).
52. Rosefield, R. US farm women: Their part in farm work and decision making. *Work Occup.* **1986**, *13*, 179–202. [CrossRef]
53. Garner, E.; De La, O.; Campos, A. *Identifying The "Family. Farm": An Informal Discussion of the Concepts and Definitions*; ESA Working Paper No. 14–10; FAO: Rome, Italy, 2014.
54. Chiswell, H.M. The importance of next generation farmers: A conceptual framework to bring the potential successor into focus. *Geogr. Compass* **2014**, *8*, 300–312. [CrossRef]
55. Cassidy, A.; McGrath, B. The relationship between 'non-successor' farm offspring and the continuity of the Irish family farm. *Sociol. Rural.* **2014**, *54*, 399–416. [CrossRef]
56. Ghaljaie, F.; Naderifar, M. Snowball sampling A purposeful method of sampling in qualitative research. *Strides Dev. Med. Educ.* **2017**, *4*, 1–4. [CrossRef]
57. Price, L. The Women's Land Army and Geographies of Gender Identity: Remembering the Impact. In Proceedings of the Association of American Geographers Annual Conference, New York, NY, USA, 1–2 February 2012; pp. 1–3. Available online: https://pure.qub.ac.uk/en/publications/the-womens-land-army-and-geographies-of-gender-identity-rememberi (accessed on 3 November 2021).
58. Alston, M. Gender-based-violence in post-disaster recovery situations: An emerging public health issue. In *Gender Based Violence and Its Impact on Health*; Nakray, K., Ed.; Routledge: Oxfordshire, UK, 2012.
59. Weeks, J. Telling stories about men. *Sociol. Rev.* **1996**, *44*, 746–756. [CrossRef]
60. Anderson, J. Talking whilst walking: A geographical archaeology of knowledge. *AREA* **2004**, *36*, 254–261. [CrossRef]
61. Riley, M.; Harvey, D. *Place-Based Interviewing: Creating and Conducting Walking Interviews*; SAGE Publications: London, UK, 2016; Available online: http://www.doi.org/10.4135/9781446273030155595386 (accessed on 10 November 2021).
62. Riley, M. Bringing the 'invisible farmer' into sharper focus: Gender relations and agricultural practices in the Peak District (UK). *Gend. Place Cult.* **2009**, *16*, 665–682. [CrossRef]
63. Riley, M. Experts in Their Fields: Farmer — Expert Knowledges and Environmentally Friendly Farming Practices. *Environ. Plan A.* **2008**, *40*, 1277–1293. [CrossRef]
64. Riley, M. Ask the Fellows Who Cut the Hay: Farm Practices, Oral History and Nature Conservation. *Oral Hist.* **2004**, *32*, 45–53.
65. Evans, J.; Jones, P. The walking interview: Methodology, mobility and place. *Appl. Geogr.* **2011**, *31*, 849–858. [CrossRef]
66. Riley, M. Emplacing the research encounter: Exploring farm life histories. *Qual. Inq.* **2010**, *16*, 651–662. [CrossRef]
67. Kuehne, G. Eight issues to think about when interviewing farmers. *Forum Qual. Social. I Forum Qual. Soc. Res.* **2016**, *17*, 20.
68. Brown, E.; Gray, R.; Monaco, S.L.; O'Donoghue, B.; Nelson, B.; Thompson, A.; Francey, S.; McGorry, P. The potential impact of COVID-19 a psychosis: A rapid review of contemporary epidemic pandemic research. *Schizophr. Res.* **2020**, *222*, 79–87. [CrossRef]
69. Lolacono, V.; Brown, P.M. Skype as a tool for qualitative research interviews. *Sociol. Res. Online* **2016**, *21*, 103–117. [CrossRef]
70. Mealer, M.; Jones, J. Methodological and ethical issues related to qualitative telephone interviews on sensitive topics. *Nurse Res.* **2018**, *21*, 32–37. [CrossRef] [PubMed]
71. Deaken, H.; Wakefield, K. Skype interviewing: Reflections of two PhD researchers. *Qual. Res.* **2013**, *14*, 603–616. [CrossRef]
72. Ajzen, I. The theory of planned behavior. *Organ. Hum. Behav. Dec. Proc.* **1991**, *50*, 121–179. [CrossRef]
73. Fenster, T. The right to the gendered city: Different formations of belonging in everyday life. *J. Gend. Stud.* **2005**, *14*, 217–231. [CrossRef]
74. Yuval-Davis, N. Belonging and the politics of belonging: Intersectional contestations. *Sociology* **2015**, *49*, 1–2. [CrossRef]
75. Maliene, V.; Dixon-Gough, R.; Malys, N. Dispersion of narrative importance values contributes to the ranking uncertainty. *Appl. Softw. Comput.* **2018**, *67*, 286–298. [CrossRef]
76. Gasson, R.; Errington, A. *The Family Farm Business*; CAB International: Wallingford, UK, 1993.
77. Tovey, H. Risk, morality and the sociology of animals—Reflections on the foot and mouth outbreak in Ireland. *Irel. J. Sociol.* **2002**, *11*, 23–42. [CrossRef]
78. Bock, B.; van Huik, M.; Prutzer, M.; Kling Eveillard, F.K.; Dockes, A. Farmers' relationship with different animals: The importance of getting close to the animals. Case studies of French, Swedish and Dutch cattle, pig and poultry farmers. *Int. J. Sociol. Food Agric.* **2007**, *15*, 108–125.
79. Riley, M. Still Being the 'Good Farmer': (Non-)retirement and the Preservation of Farming Identities in Older Age. *Sociologia Ruralis* **2016**, *56*, 96–115. [CrossRef]
80. DAERA. Available from: About DAERA Department of Agriculture, Environment and Rural Affairs (daera-ni.gov.uk). Available online: http://www.daera-ni.gov.uk/ (accessed on 28 November 2020).
81. Conway, S.F.; McDonagh, J.; Farrell, M.; Kinsella, A. Till death do us part: Exploring the Irish farmer-farm relationship in later life through the lens of 'Insideness'. *Int. J. Agric. Manag.* **2018**, *7*, 1–13.

MDPI

Commentary

'Farmers Don't Retire': Re-Evaluating How We Engage with and Understand the 'Older' Farmer's Perspective

Shane Francis Conway [1,*], Maura Farrell [1], John McDonagh [1] and Anne Kinsella [2]

1 Rural Studies Centre, Discipline of Geography, School of Geography, Archaeology and Irish Studies, National University of Ireland Galway, H91 TX33 Galway, Ireland; maura.farrell@nuigalway.ie (M.F.); john.mcdonagh@nuigalway.ie (J.M.)
2 Agricultural Economics and Farms Surveys Department, Mellows Campus, Teagasc Co., H65 R718 Athenry, Ireland; anne.kinsella@teagasc.ie
* Correspondence: shane.conway@nuigalway.ie

Abstract: Globally, policy aimed at stimulating generational renewal in agriculture is reported to pay meagre regard to the mental health and wellbeing of an older farmer, overlooking their identity and social circles, which are inextricably intertwined with their occupation and farm. This paper, in probing this contentious issue, casts its net across what could be deemed as disparate literatures, namely connected to transferring the family farm and social gerontology, in order to determine what steps could be taken to reassure older farmers that their sense of purpose and legitimate social connectedness within the farming community will not be jeopardised upon handing over the farm business to the next generation. A number of practical 'farmer-sensitive' actions that can be taken at both policy and societal level are subsequently set forth in this paper to help ease the fear and anxiety associated with 'stepping aside' and retirement from farming amongst older farmers. A particular focus is placed on social and emotional wellbeing benefits of being a member of a social group reflecting farmer-relevant values and aspirations in later life. The potential of the multi-actor EIP-AGRI initiative and the long-established livestock mart sector in facilitating the successful rollout of a social organisation designed to fit the specific needs and interests of the older generation of the farming community is then outlined. In performing this, the paper begins a broad international conversation on the potential of transforming farming into an age-friendly sector of society, in line with the World Health Organization's (WHO) age-friendly environments concept.

Keywords: family farm; older farmers; retirement; succession; wellbeing; identity; social gerontology; age-friendly environments; innovation; rural sustainability

Citation: Conway, S.F.; Farrell, M.; McDonagh, J.; Kinsella, A. 'Farmers Don't Retire': Re-Evaluating How We Engage with and Understand the 'Older' Farmer's Perspective. *Sustainability* **2022**, *14*, 2533. https://doi.org/10.3390/su14052533

Academic Editor: Marc A. Rosen

Received: 17 November 2021
Accepted: 16 February 2022
Published: 22 February 2022

Publisher's Note: MDPI stays neutral with regard to jurisdictional claims in published maps and institutional affiliations.

1. Introduction

Global demographic trends highlight an inversion of the age pyramid with those aged 65 years and over, constituting the fastest growing sector of the farming community [1,2]. This 'greying' of the farming workforce necessitates an infusion of 'new blood' into the agricultural industry by means of efficient and effective intergenerational farm succession and land mobility (i.e., transfer of land from one farmer to another or from one generation to the next) in order to ensure future prosperity of the farming sector, as well as long-term sustainability of food production systems [3,4].

Extensive research on social and emotional issues affecting older farmers [5–8] highlights, however, their overwhelming desire to remain actively engaged in farming in later life as it is central to their sense of self and belonging in the farming community. The prospect of surrendering professional and personal identity upon transferring managerial control of the farm to the next generation and in some cases retiring, as advised in Europe's Common Agricultural Policy (CAP), for example, is a concept that older farmers find difficult to accept. In fact, there appears to be a cultural expectation within the farming community that 'farmers don't retire' [5]. Lobley et al. added that the most common approach to farm retirement may not actually be retirement per se but rather remaining in situ and 'continuing day-to-day involvement' on the farm, albeit with a reduction in some of the 'more arduous tasks' [9] (p. 51). In addition, farmers often wish to remain 'rooted in place' on the farm and, in many cases, have developed few interests outside of farming, further reinforcing their reluctance to step away from farming in later life [10]. Price and Evans go so far as to describe farmers as being so deeply embedded in the 'cultural and physical spaces of farming', that they 'cannot imagine a different way of life' [11] (p. 6). For those farmers who have retired, challenges are also present, with Riley [12] identifying adverse effects that the transition process has had on their lives. Riley explained that the cessation of occupational engagement 'not only left voids in terms of time and empty routine structures, but also the loss of a lens through which they channelled very particular understandings of, and relationships with, specific places and practices' [12] (p. 23). Riley further added that retirees felt 'lost' upon ceasing their 'association with, and everyday routines and actions within', the farm space [13] (p. 770).

The challenges that this presents are not insignificant. Globally, it could be argued that generational renewals in agriculture policymakers, and key stakeholders, are preoccupied with developing strategies and interventions to encourage older farmers to 'step aside' to facilitate young farmers who want to establish a career in farming. Such strategies, however, pay scant regard to the mental health and wellbeing of senior generations ignoring as it does their identity and social circles which are inextricably intertwined with their occupation and farm [14,15]. This paper, in probing this global issue, casts its net across what could be deemed as disparate literatures (transferring the family farm and social gerontology) in order to determine what, if any, steps could or should be taken to minimise the disconnect between policy and realities on the ground. In particular, the paper focuses on practical actions that can be taken at both policy and societal levels to reassure older farmers that their sense of purpose and legitimate social connectedness within the farming community will not be jeopardised upon handing over the farm business to the next generation. In performing this, we begin a broad international conversation on the place of older farmers in society in general and the part this plays in generational renewal and broader agriculture policy narratives.

This paper, by interrogating agricultural policy responses to an ageing farming population, considers potential strategies and interventions to help ease the fear, anxiety and stress associated with intergenerational farm succession amongst the older generation of the farming community. A particular focus is placed on the wellbeing benefits of being a member of a social group reflecting farmer relevant values and aspirations in later life. The extent to which older farmers themselves can be involved in the coproduction of a comprehensive set of 'farmer-sensitive' generational renewal policies and practices at the farm level, which respond to their needs and concerns in later life, will also be explored. This process, we argue, can have a global reach and could be extended in various ways to other farming communities.

This insight is particularly timely in the current COVID-19 pandemic as rural communities prepare to adapt, rebuild and reenergize as part of their crisis recovery plans. Social isolation measures brought into effect in an effort to curb the spread of the virus have further highlighted the importance of ensuring social inclusion for the elderly population (including older farmers) of society [16,17]. Indeed, the United Nations [18] has recently highlighted how support networks, such as retirement groups, played a valuable and beneficial role in ensuring and securing the wellbeing of older people during these difficult times.

The next section reviews pertinent family farm transfer and social gerontology literature published over the past forty years. Connecting these fields of study is paramount in fulfilling the overall purpose of this paper, as a wealth of peer-reviewed social gerontology journal articles and edited book chapters exists on age-related changes in social activity and engagement in later life, whereas its counterpart from a farming perspective is largely absent. This is followed by a series of agri-social policy recommendations aimed at addressing the older generation of the farming community's social and emotional wellbeing and what this might look like in any future vision for rural society.

2. Literature Review

2.1. Greying' of the Farming Workforce

The 'greying' of the farming workforce is a pressing issue throughout the world. In Europe, for example, almost one-third of all farmers are older than 65 [2]. This raises obvious concerns about the survival, continuity and future of the agricultural industry as well as the broader sustainability of the rural society [1,7]. The situation in the Republic of Ireland is closely analogous to that of its European counterparts; in 2016, 30% of Irish farm holders were under 65 years and over [19]. Similarly, in the USA, the most recent Census of Agriculture carried out in 2017 found that 34% of farmers are aged 65 and older [20]. Indeed, this census also highlights that the average age of farmers in the USA is now 57.5 years, up 1.2 years from 2012, continuing a long-term trend of ageing in the U.S. farming population (ibid).

The senior generation's reluctance and indeed resistance to alter the status quo of the existing management and ownership structure of the family farm, to help facilitate generational renewal in agriculture, is undoubtedly strong within the farming community [7,21]. The reasons why older farmers fail to plan, effectively and expeditiously, for the future are expansive and range from the potential loss of identity, status and power that may occur as a result of engaging in the process to the intrinsic, multilevel relationship farmers have with their farms. The common denominator, however, is that intergenerational farm transfer is about emotion. The so-called 'soft issues', i.e., the emotional and social issues [6], are the issues that distort and dominate the older generation's decisions on the future trajectory of the farm. Such issues have resulted in intractable challenges for farm succession and retirement policy over the past fifty years. As such, far from being the 'soft issues', these really are the 'hard issues'. Similar findings from a collaborative research effort called the International FARMTRANSFERS Project, yielding a range of (largely quantitative) data relating to the pattern; process and speed of farm succession; retirement and inheritance that has been undertaken in 11 countries and 7 states in the USA and completed by almost 16,000 farmers throughout the world, highlights the global scale of such concerns [22]. Farming is a way of life for many older farmers throughout the world and there can be detrimental consequences to their emotional wellbeing if they are 'cut off' from their daily routines on the farm and social circles in the farming community [22–24]. Riley suggests that the 'indivisibility of social and occupational spaces' within the farming community leaves farmers feeling isolated or like 'an outsider' within previously 'familiar and comfortable spaces' following retirement [13] (p. 769). Such concerns are more prevalent than ever in the context of increased levels of social isolation and loneliness in rural communities due to the COVID-19 pandemic.

2.2. Social Isolation in Later Life

It is widely acknowledged in social gerontology research that social isolation and withdrawal from active engagement is a common occurrence in later life [25–27]. Social isolation is defined as an absence or shortage of social interaction and connections between an individual and the rest of society [28]. Disengagement from society in later life largely occurs due to a decrease in one's social participation or loss of identity upon ceasing formal employment (ibid). Previous research carried out by Berg and Cassells highlight the negative impacts social isolation can have on the older population's psychological functioning as they struggle to cope with changes in their lives, particularly post-retirement [29]. This major life transition can potentially result in significant challenges to an individual's health and wellbeing due to its association with increased isolation and loneliness [30]. Older people in rural areas are reported to be the most vulnerable, at-risk segment of the population with regards to social isolation, as they generally have smaller social networks and are more likely to be living by themselves [16,31,32]. Within this subset of older people in rural areas, we have the older farming population and the particular challenges they face and indeed the possibilities they can offer by connecting to a social environment that reflects their own values and aspirations [30].

2.3. Active Social Engagement

Social gerontologists and psychologists have been exploring the relationship between social activity, life satisfaction and healthy ageing since the 1970s. Lemon et al., for example, explained that the social self emerges and is sustained in a most basic manner through interaction with others and structural constraints, which limit or deny contacts with the environment and tend to be alienating and demoralizing [33]. Rowe and Kahn highlighted that continued social participation is one of three major elements of successful ageing, along with a low prevalence of illness and high physical and cognitive capabilities [34]. Rowe and Kahn also added that active social engagement in later life involves maintaining close relationships as well as remaining involved in activities that are meaningful and purposeful [34]. This ideal very much reflects the challenges facing the older farming community. Social relationships are observed to help negate negative feelings, with knock-on effects for physical health in the longer term, particularly within rural communities [35]. Indeed, social integration and inclusion for older people can have a profound impact on one's psychological wellbeing in later life [36,37].

2.4. The Importance of a Social Group in Later Life

The fundamental role that membership of a social group can play in protecting the physical and mental wellbeing of the older generation is well documented in social gerontology research [35,36,38,39]. Steffens et al. (2016) explained how older people derive a sense of self-esteem, purpose, acceptance and belonging in later life if they participate in broader social groups. The WHO add that becoming a member of a social group, and the subsequent meaningful activities that materialise upon doing so in a social context, can enhance one's quality of life in old age [40]. Greaves and Farbus and Crabtree et al. also highlighted that membership of a locally integrated social network helps combat and reduce social isolation amongst the older generation by providing them with an outlet to reconnect with their community in later life [36,41]. Spurgin added that membership in social groups was especially important for older people adapting to significant and anxiety provoking life transitions such as retirement [42]. Steffens et al. explained that being a member of a social group post-retirement not only has the potential to increase an older person's overall quality of life but also may increase life expectancy, due to the profound benefits they hold for self-identity, self-esteem, resilience and mental health [30]. The effectiveness of social groups and active retirement organisations in enhancing social networks and improving the psychological and physical wellbeing of older people is also widely reported in an Irish context [35,43]. One such successful social group in existence in Ireland is Men's Sheds, an organisation which first emerged in Australia in the 1990s as a response to the increasing concerns about men's health [44]. Crabtree et al. found that Men's Sheds not only improved older men's levels of social interaction but also enhanced their optimism and willingness to make a positive difference to their own lives [44]. Ni Leime et al. added that membership of a social group in later life also encourages older people to engage in activities that they would not have otherwise engaged in, adding a new dimension to their lives [43].

2.5. Disconnected Policy Efforts

Evidence from social gerontology research regarding the mental health and wellbeing benefits of continued social inclusion and participation in later life brings to the fore the risks associated with strategies and interventions aimed at facilitating intergenerational farm transfer, such as the most recent largely unsuccessful Early Retirement Scheme for farmers in Ireland (ERS 3), which required farmers to intend to retire under a scheme to 'cease agricultural activity forever' [45]. Such a sentiment fails to take into account the emotional and social welfare of older farmers. Indeed, a recent report by the European Commission evaluating the impact of the CAP on generational renewal, local development and jobs in rural areas criticised the appropriateness of such Early Retirement Scheme (ERS) measures and called for an increased focus on mechanisms that help older farmers enhance their 'quality of life by exploring possibilities under social policy' [46] (p. 48). Such recommendations came almost forty years after the late Dr. Patrick (Packie) Commins first stressed that farm retirement policy should not focus on 'economic objectives' alone and should 'not ignore possible social consequences or wider issues of human welfare' in the early 1970s, however [47].

It is now imperative that policymakers crafting generational renewal strategies, and key stakeholders, finally recognise and appreciate the enormous value added to older farmer's personal lives and social circles through active engagement in farming in later life. Taking this route, the recommendations set forth in the next section will not only help address the intergenerational farm succession difficulties but will also protect the health and wellbeing of older farmers in the context of their own rural future.

3. Recommendations

In the face of such academic and popular support for active social engagement and social membership in later life within social gerontology research, it is surprising, therefore, that there are no social groups currently in existence in the agricultural sector that specifically represent the needs and interests of the older farmers. This paper recommends the establishment of a national social organisation, akin to that in place for young farmers in rural Ireland, i.e., Macra na Feirme, for the older generation of the farming community. Such an organisation would be a hugely positive step forward and could be a key method for policy to respond positively to the ageing farming population. In the following sections, focusing solely on the Irish example, we outline how such an organisation in a farming context might be set up and ultimately the role it could play in protecting the mental health and wellbeing of the older generation of the farming community. This we believe is particularly timely in an era of increased international attention on rural sustainability and the necessity of stimulating generational renewal in the farming sector through intergenerational farm succession and retirement.

3.1. Establishment of a National Social Organisation for Older Farmers

From an Irish viewpoint, a national social organisation for older farmers, funded annually by the Government and through membership, would provide older farmers with a sense of purpose and legitimate social connectedness within the farming community, thus helping to ease anxiety and stresses associated with the intergenerational farm succession process. Furthermore, involvement in such an organisation would help alleviate concerns around the fear of the unknown upon doing so by having an outlet to remain embedded 'inside' the agricultural sphere in later life. A nationwide social organisation, with a network of clubs in every county (or similar geographic entity), would also promote social inclusion in farming by allowing older farmers to integrate within the social fabric of a local age peer group (Macra na Feirme in Ireland have approximately 200 clubs in 31 regions). Membership of such a group would provide opportunities to develop a pattern of farming activities suited to advancing age through increased collaboration with farmers at a similar stage of their lives. This would contribute to an overall sense of happiness and self-worth, amidst the gradual diminishment of physical capacities on the farm. Such peer-to-peer collegiality and comradeship would be particularly beneficial for farmers living alone or for those who do not have successor in situ to take over the farm.

Despite the widely reported successful health and wellbeing benefits of the aforementioned Men's Sheds movement for older men [41], this paper advises that this proposed national social organisation for older farmers would be open to all within the farming community, thus helping dismantle the patriarchal nature of family farming identified in previous research [48–51]. The gender diversity of such a social group would also help bring about increased levels of social inclusion amongst older women involved in farming.

Similarly to Macra na Feirme, this body for older farmers, with their added wealth of experience, would also act as a social partner farm organisation. The organisation in turn could collaborate with the Irish Farmers Association (IFA), for example, which would allow them to have regular access to government ministers and senior civil servants, thus providing older farmers with a voice to raise issues of concern through active participation and engagement with stakeholder groups. Indeed, such a group could be invaluable with regard to the development of future farm transfer strategies that would be cognisant of the human side of the process of generational renewal for example. An established organisation for older farmers would allow this sector of society to have a representative on important committees, such as the Board of Teagasc (similar to their younger counterparts). Such a feeling of belonging, inclusivity and position in society will undoubtedly contribute to the senior generation's general satisfaction and wellbeing in later life. A plethora of social gerontology studies indicate that activities and interventions, which promote active engagement, encourage creativity and facilitate social contact, have positive effects on one's health and wellbeing in later life, with knock-on effects for physical health in the longer term [36,43,52].

Collaborating with younger counterparts in Macra na Feirme on campaigns and activities through such a social organisation would help the senior generation of the farming community retain a sense of purpose and value. Consequently, the younger generation of farmers can learn from the older generation's invaluable store of local knowledge developed over years of regularised interaction and experience working in the farming sector. Such 'soil-specific human capital', as it is often referred to [53], is not easily transferable, communicated or learnable. It is highly plausible, for example, that the criteria of the recent Early Retirement Scheme in Ireland (ERS 3), requesting that 'continued participation in farming is not permitted' [38], had a profoundly negative effect on farm performance as it resulted in the loss of a number of experienced personnel from the agricultural sector. A national social organisation for older farmers has the potential to address such failings by facilitating the intergenerational exchange of agricultural knowledge and skills in a collaborative manner. It is a win-win situation in that such contributions would support a more viable, sustainable and vibrant agri-food and rural sector.

This proposed social organisation for older farmers would also, we argue, have the potential to create an age-friendly environment in the farming sector. The concept of age-friendly environments has garnered international attention among researchers, policy makers and community organisations since the World Health Organization (WHO) launched its Global Age-friendly Cities and Communities project in 2006. Although there is no universally accepted definition of an 'age-friendly' environment, the WHO defines an age-friendly community as one in which 'policies, services, settings and structures support and enable people to age actively' [54] (p. 5). Despite the growth of the age-friendly environments movement, existing literature is predominantly focused on a model of urban ageing that fails to reflect the broader diversity of rural areas and more significantly that of the farming community. The establishment of a social organisation for older farmers can help address this significant underrepresentation by generating a culture of appreciation and respect for their way of life, both within policy circles and society more generally. The rollout of such a group, while from initial viewing may seem challenging, upon closer inspection could be relatively easy to instigate. To this end, we would propose drawing on already existing conduits, namely that of the EIP-AGRI Operational Group Projects and the long-established livestock mart sector.

3.2. Rolling out a National Social Organisation for Older Farmers

Step 1: Formation of an EIP-AGRI Operational Group

The formation of an EIP-AGRI (European Innovation Partnership for Agricultural productivity and Sustainability) Operational Group, designed to fit the specific needs and concerns of the senior generation of the farming community, would be a positive step in this process. A pilot study via the EIP-AGRI format would be an ideal stepping stone for the full implementation of a national social organisation for older farmers. The EIP-AGRI initiative was launched by the European Commission's Directorate-General for Agriculture and Rural Development's (DG-AGRI) in 2012 to contribute to the European Union's strategy 'Europe 2020' for smart, sustainable and inclusive growth [55]. As innovation under the EIP-AGRI initiative can be social as well as technological and nontechnological, such a multi-actor EIP-AGRI project led by a group of older farmers, in collaboration with farm advisory bodies, research institutions and organisations such as Age Action Ireland, has the potential to generate a comprehensive set of age-friendly policies and practices at the farm level. Furthermore, such an Operational Group has the potential to bring about much needed change in mindset and mannerisms of the farming community towards a range of contemporary concerns. This includes issues of land mobility and farm succession, but it would also help provide older farmers with a solid platform to become directly involved in the development and implementation of more 'farmer-sensitive' policies aimed at stimulating generational renewal in agriculture. Such measures would ultimately help bring about a culture within the farming community of recognising the importance of engaging in the farm succession planning process in a timely manner [7].

What this could also provide in terms of associated positive spin-off is the potential it would also have in addressing a significant challenge, namely that of health and safety on the farm. Farming is one of the most hazardous occupations in terms of the incidence and seriousness of accidental injuries [56]. Moreover, agriculture exhibits disproportionately high fatality rates, when compared to other sectors [57]. With 66% of farm fatalities in Ireland involving farmers aged 60 and over in 2019 for example [58], this phenomenon requires immediate policy intervention. Many are unwilling to recognize or accept their physical limitations on the farm [59] and, instead, continue to traverse spaces that would appear to be beyond their level of physiological competence [60], with subsequent risks to their health and safety. The general satisfaction and wellbeing that the older generation of the farming community attribute to the daily and seasonal labour-intensive demands of working on the farm in later life appears to be part of the farming psyche. As such, the establishment of a national social organisation for older farmers would provide the Health and Safety Authority (HSA) and member organisations of the HSA Farm Safety Partnership Advisory committee in the Republic of Ireland with an invaluable platform to directly engage with older farmers on the various actions to handle age-related physical limitations and barriers on their farms. The in-depth, farmer-focused, insights into the intrinsic link to farm attachment in old age and the importance attributed to the habitual routines within the farm setting made possible by interacting and collaborating with such an organisation for older farmers would aid HSA in the development of an effective health and safety service tailored specifically to the needs of older farmers.

Furthermore, the European Commission's Directorate-General for Agriculture and Rural Development's AKIS (Agriculture Knowledge and Innovation Systems) ecosystem acknowledges and recognises the need for EU Member States to include social innovation and inclusiveness within their AKIS knowledge exchange strategies and action plans [61]. Taking into account the range of rural socio-cultural contexts in their respective countries, this EIP-AGRI project for older farmers, tuned to the human side of farming in later life, will illustrate how policy, and indeed society more generally, can respond positively to the ageing farming population.

Step 2: Collaborating with the Livestock Mart Sector

A second stage in the formation of this new group, could stem from the livestock mart sector. The value attached to the social dimension of attending a livestock market is widely reported in the United Kingdom [62,63], and in Ireland, this sector consists of over 60 cooperative mart centres across the Irish countryside. These marts, we suggest, have the capacity and scale to help roll out a national social organisation for older farmers, albeit with some additional government supports that are probably necessary. In addition to their primary function providing a consistent, stable and transparent method of buying and selling livestock through a guaranteed payment structure, marts also provide a vital social facility for the farming community, some of whom have no other social outlet [63]. Indeed, many older farmers rely on their weekly visit to the mart to meet friends, exchange ideas and to catch up on local news in an informal setting. This has almost grown in significance in recent years as many of the natural meeting points within rural communities have been removed due to the closure of post offices, pubs and local shops for example [64]. The bidding ring and canteen at livestock marts are, therefore, not considered to be only venues of transaction within the farming community but constitute hives of social interaction. This paper, therefore, proposes that livestock marts could be drawn on to help facilitate a national organisation as they have a considerable role already in terms of providing a social hub for the older generation of the farming community. Their existing positionality and reputation as a 'buzz' of activity within the heart of rural communities provide livestock marts with a ready-made platform and network to establish a social group membership of older farmers in their catchment area.

4. Conclusions

Overcoming the farming community's stalwart persistence in their adherence to traditional succession and retirement practices evident in previous family farm transfer research, which effectively obstruct the transfer of farmland from one generation to the next, is a pressing matter for contemporary generational renewal policy. This is not confined to one country but has a global dimension. Nonetheless, there is limited focus on the place of the older farmer within policy/academic discussion, even though this cohort ultimately has the resources and power to decide whether the transition takes place or not [21,65].

Consequently, there is an urgent need for agriculture policy makers and practitioners to re-examine their existing predominant focus on addressing the needs and requirements of the younger farming generation and to place a greater or equal emphasis on maintaining the quality of life of those most affected by the process, namely the older farmer. The recommendations set forth in this paper are, therefore, aimed at addressing older farmers' emotional and social wellbeing, as well as their sense of purpose in later life. We argue here for the establishment of a national social organisation for older farmers which would have the potential to transform farming into an age-friendly sector of society. Increasing evidence within social gerontology research on the benefits and importance of such social organisations on the lives of older people, particularly in relation to ensuring social inclusion in later years, reaffirms why there should be immediate support of such a venture in the farming sector. Making use of the extensive reach of the multi-actor EIP-AGRI initiative, allied to the platform that Ireland's livestock mart sector presents, has the potential to rollout such a national social organisation for the older generation of the farming community in an effective and efficient manner. Whilst our recommendations here are predominately geared towards influencing policy in the Republic of Ireland and its impact and success may be dependent on the cultural and institutional milieu that govern the mindset and mannerisms of Irish farmers in later life, this paper is also applicable in a much broader European and global settings.

While the establishment of such a social organisation for older farmers could be viewed as unnecessary and/or too idealistic, we need to remember that we all inevitably have to face the prospect of letting go of our professional tasks and ties in our old age. No one can avoid ageing, and as extensive research on the human dynamics affecting farm succession and retirement over the past decade has identified, the majority of older farmers opt to remain actively engaged in farming in later life in order to help ensure legitimate social connectedness within the farming community. The formation of a national social organisation for older farmers would provide older farmers with an outlet to engage with their peers and develop a pattern of farming activities suited to advancing age. Furthermore, it would offer the older generation of the farming community an ideal platform to become involved in the coproduction of a comprehensive set of age-friendly policies and practices at farm level, which respond to their aspirations and needs in later life. Such insight and input will help inform future generational renewal in agriculture policy directions that understand the world as older farmers perceive it, and consequently prevent them from becoming isolated and excluded from society almost by accident rather than intention. Aligning such policies to the World Health Organization's (WHO) age-friendly environments concept will also induce a much broader international conversation on the place, views, concerns and challenges of older farmers in the context of the future viable of the agricultural sector and ultimately the future sustainability of rural families, communities and natural environments upon which we all depend.

Author Contributions: Conceptualization, S.F.C., M.F. and J.M.; writing—review and editing, S.F.C., M.F., J.M. and A.K.; funding acquisition, S.F.C. All authors contributed to writing, editing and reviewing the draft submitted. All authors have read and agreed to the published version of the manuscript.

Funding: This study was funded by the NUI Galway College of Arts, Social Sciences and Celtic Studies Illuminate Programme.

Institutional Review Board Statement: Not applicable.

Informed Consent Statement: Not applicable.

Data Availability Statement: Not applicable.

Acknowledgments: The lead author would like to thank the Innovation Office at NUI Galway for their assistance with this study.

Conflicts of Interest: The authors declare no conflict of interest. The funders had no role in the design of the review; the writing of the manuscript; or in the decision to publish the results.

References

1. Dwyer, J.; Micha, E.; Kubinakova, K.; Van Bunnen, P.; Schuh, B.; Maucorps, A.; Mantino, F. *Evaluation of the Impact of the CAP on Generational Renewal, Local Development and Jobs in Rural Areas*; Technical Report; European Commission: Brussels, Belgium, 2019.
2. Eurostat. *Agriculture Statistics at Regional Level, Statistics Explained*; Eurostat: Luxembourg, 2020.
3. Zagata, L.; Sutherland, L. A Deconstructing the 'young farmer problem in Europe': Towards a research agenda. *J. Rural Stud.* **2015**, *38*, 39–51. [CrossRef]
4. McDonagh, J.; Farrell, M.; Conway, S.F. The Role of Small-scale Farms and Food Security. In *Sustainability Challenges in the Agrofood Sector*; Bhat, R., Ed.; Wiley Blackwell: Hoboken, NJ, USA, 2017; pp. 33–47.
5. Conway, S.F.; McDonagh, J.; Farrell, M.; Kinsella, A. Cease agricultural activity forever? Underestimating the importance of symbolic capital. *J. Rural Stud.* **2016**, *44*, 164–176. [CrossRef]
6. Conway, S.F.; McDonagh, J.; Farrell, M.; Kinsella, A. Human dynamics and the intergenerational farm transfer process in later life: A roadmap for future generational renewal in agriculture policy. *Int. J. Agric. Manag.* **2019**, *8*, 22–30.
7. Conway, S.F.; Farrell, M.; McDonagh, J.; Kinsella, A. Going Against the Grain: Unravelling the Habitus of Older Farmers to Help Facilitate Generational Renewal in Agriculture. *Sociol. Rural.* **2021**, *61*, 602–622. [CrossRef]
8. Holloway, L.A.; Catney, G.; Stockdale, A.; Nelson, R. Family Farming Futures: Exploring the Challenges of Family Farm Decision Making through an Emotional Lens of 'Belonging'. *Sustainability* **2021**, *13*, 12271. [CrossRef]
9. Lobley, M.; Baker, J.R.; Whitehead, I. Farm succession and retirement: Some international comparisons. *J. Agric. Food Syst. Community Dev.* **2010**, *1*, 49–64. [CrossRef]
10. Conway, S.F.; McDonagh, J.; Farrell, M.; Kinsella, A. Till death do us part: Exploring the Irish farmer-farm relationship in later life through the lens of 'Insideness'. *Int. J. Agric. Manag.* **2018**, *7*, 3–15.
11. Price, L.; Conn, R. 'Keeping the name on the land': Patrilineal succession in Northern Irish family farming. In *Keeping it in the Family: International Perspectives on Succession and Retirement on Family Farms*; Lobley, M., Baker, J., Whitehead, I., Eds.; Ashgate: Surrey, UK, 2012; pp. 93–110.
12. Riley, M. 'Letting them go'—agricultural retirement and human-livestock relations. *Geoforum* **2011**, *42*, 16–27. [CrossRef]
13. Riley, M. 'Moving on'? Exploring the geographies of retirement adjustment amongst farming couples. *Soc. Cult. Geogr.* **2012**, *13*, 759–781. [CrossRef]
14. Convery, I.; Bailey, C.; Mort, M.; Baxter, J. Death in the wrong place? Emotional geographies of the UK 2001 foot and mouth disease epidemic. *J. Rural Stud.* **2005**, *21*, 99–109. [CrossRef]
15. Rogers, M.; Barr, N.; O'Callaghan, Z.; Brumby, S.; Warburton, J. Healthy ageing: Farming into the twilight. *Rural Soc. Work Environ.* **2013**, *22*, 251–262. [CrossRef]

16. Berg-Weger, M.; Morley, J.E. Loneliness and social isolation in older adults during the Covid-19 pandemic: Implications for gerontological social work. *J. Nutr. Health Aging* **2020**, *24*, 456–458. [CrossRef] [PubMed]
17. OECD. *COVID-19: Protecting People and Societies, Policy Responses to Coronavirus (COVID-19)*; OECD: Paris, France, 2020.
18. United Nations. *The Impact of COVID-19 on Older Persons, Policy Brief, May 2020*; United Nations: New York, NY, USA, 2020.
19. Central Statistics Office (CSO). *Statistical Yearbook of Ireland 2018*; Farms and Farmers; Central Statistics Office: Dublin, Ireland, 2018.
20. USDA (United States Department of Agriculture National Agricultural Statistics Service). *2017 Census of Agriculture Highlights*; ACH17-2/April 2019; National Agricultural Statistics Service: Washington, DC, USA, 2019.
21. Conway, S.F.; McDonagh, J.; Farrell, M.; Kinsella, A. Uncovering Obstacles: The Exercise of Symbolic Power in the Complex Arena of Intergenerational Family Farm Transfer. *J. Rural. Stud.* **2017**, *54*, 60–75. [CrossRef]
22. Lobley, M.; Baker, J. Succession and retirement in family farm businesses. In *Keeping It in the Family: International Perspectives on Succession and Retirement on Family Farms (Ashgate)*; Lobley, M., Baker, J., Whitehead, I., Eds.; Ashgate: Surrey, UK, 2012; pp. 1–19.
23. Barclay, E.; Reeve, I.; Foskey, R. Australian farmers' attitudes towards succession and inheritance. In *Keeping it in the Family: International Perspectives on Succession and Retirement on Family Farms (Ashgate)*; Lobley, M., Baker, J.R., Whitehead, I., Eds.; Ashgate: Surrey, UK, 2012; pp. 21–36.
24. Kuehne, G. My decision to sell the family farm. *Agric. Hum. Values* **2013**, *30*, 203–213. [CrossRef]
25. Srivastava, K.; Das, R.C. Personality pathways of successful ageing. *Ind. Psychiatry J.* **2013**, *22*, 1–3. [CrossRef]
26. Department of Health (DoH). *Positive Ageing—Starts Now! The National Positive Ageing Strategy (NPAS)*; Department of Health (DoH): Dublin, Ireland, 2018.
27. Ward, M. *Irish Adults' Transition to Retirement—Wellbeing, Social Participation and Health-Related Behaviours*; The Irish Longitudinal Study on Ageing (TILDA): Dublin, Ireland, 2019.
28. Xia, N.; Li, H. Loneliness, social isolation, and cardiovascular health. *Antioxid. Redox Signal.* **2018**, *28*, 837–851. [CrossRef]
29. Berg, R.; Cassells, J. *The Second Fifty Years: Promoting Health and Preventing Disability. Institute of Medicine (US) Division of Health Promotion and Disease Prevention*; National Academy Press: Washington, DC, USA, 1992.
30. Steffens, N.K.; Cruwys, T.; Haslam, C.; Jetten, J.; Haslam, A.S. Social group memberships in retirement are associated with reduced risk of premature death: Evidence from a longitudinal cohort study. *BMJ Open* **2016**, *6*, e010164. [CrossRef]
31. Henning-Smith, C.; Moscovice, I.; Kozhimannil, K. Differences in social isolation and its relationship to health by rurality. *J. Rural Health* **2019**, *35*, 540–549. [CrossRef]
32. Seyfzadeh, A.; Haghighatian, M.; Mohajerani, A. Social isolation in the elderly: The neglected issue. *Iran. J. Public Health* **2019**, *48*, 365–366. [CrossRef]
33. Lemon, B.W. Bengtson, V.L.; Peterson, J.A. An Exploration of the Activity Theory of Ageing: Activity Types and Life-Satisfaction among In-Movers to a Retirement Community. *J. Gerontol.* **1972**, *27*, 511–523. [CrossRef]
34. Rowe, J.W.; Kahn, R.L. Successful aging. *Gerontologist* **1997**, *37*, 433–440. [CrossRef] [PubMed]
35. Dorgan, M. *Exploring the Needs for Social Inclusion in Rural IRELAND: Ballinora a Case Study*; Community-Academic Research Links, University College Cork: Cork, Ireland, 2013.
36. Greaves, C.J.; Farbus, L. Effects of creative and social activity on the health and wellbeing of socially isolated older people: Outcomes from a multi-method observational study. *J. R. Soc. Promot. Health* **2006**, *126*, 134–142. [CrossRef] [PubMed]
37. Gallagher, C. Connectedness in the lives of older people in Ireland: A study of the communal participation of older people in two geographic localities. *Ir. J. Sociol.* **2012**, *20*, 84–102. [CrossRef]
38. Desrosiers, J.; Gosselin, S.; Lelclerc, G.; Gaulin, P.; Trottier, L. Impact of a specific program focusing on the psychological autonomy and actualization of potential of residents in a long-term care unit. *Arch. Gerontol. Geriatr.* **2000**, *31*, 133–146. [CrossRef]
39. Gyasi, R.M.; Phillips, D.R.; Abass, K. Social support networks and psychological wellbeing in community-dwelling older Ghanaian cohorts. *Int. Psychogeriatr.* **2019**, *31*, 1047–1057. [CrossRef]
40. World Health Organization. *The Global Strategy and Action Plan on Ageing and Health*; World Health Organization: Geneva, Switzerland, 2017.
41. Crabtree, L.; Tinker, A.; Glaser, K. Men's sheds: The perceived health and wellbeing benefits. *Work. Older People* **2018**, *22*, 101–110. [CrossRef]
42. Spurgin, L. Attending Social Groups Can Improve Health and Wellbeing after Retirement. 2016. Available online: https://www.thereader.org.uk/attending-social-groups-can-improve-health-and-wellbeing-after-retirement/ (accessed on 1 September 2021).
43. Ni Leime, A.; Callan, A.; Finn, C.; Healy, R. *Evaluating the Impact of Membership of ACTIVE Retirement Ireland on the Lives of Older People*; Active Retirement Ireland: Dublin, Ireland, 2012.
44. Kelly, D.; Steiner, A.; Mason, H.; Teasdale, S. Men's Sheds: A conceptual exploration of the causal pathways for health and well-being. *Health Soc. Care Community* **2019**, *27*, 1147–1157. [CrossRef]
45. DAFM (Department of Agriculture, Food and the Marine). *Terms and Conditions for Early Retirement Scheme (ERS3)*; DAFM: Dublin, Ireland, 2007.
46. European Commission. *Evaluation of the Impact of the CAP on Generational Renewal, Local Development and Jobs in Rural Areas {SWD(2021) 79 Final}*; Technical Report; European Commission: Brussels, Belgium, 2021.
47. Commins, P. *Retirement in Agriculture: A Pilot Survey of Farmers' Reactions to E.E.C. Pension Schemes*; Macra na Feirme: Dublin, Ireland, 1973.
48. Gasson, R.; Errington, A. *The Farm Family Business*; CAB-International: Wallingford, CT, USA, 1993.

49. Brandth, B. Gender Identity in European Family Farming. *Sociol. Rural.* **2002**, *42*, 181–200. [CrossRef]
50. Price, L.; Evans, N. From stress to distress: Conceptualizing the British family farming patriarchal way of life. *J. Rural Stud.* **2009**, *25*, 1–11. [CrossRef]
51. Shortall, S. Changing gender roles in Irish farm households: Continuity and change. *Ir. Geogr.* **2018**, *50*, 175–191.
52. United Nations. *Report of the Second World Assembly on Ageing: Madrid 8–12, 2002*; United Nations: New York, NY, USA, 2002.
53. Laband, D.N.; Lentz, B.F. Occupational Inheritance in Agriculture. *Am. J. Agric. Econ.* **1983**, *36*, 311–314. [CrossRef]
54. World Health Organization. *Health and Development through Physical Activity and Sport*; World Health Organization: Geneva, Switzerland, 2003.
55. EIP-AGRI. *EIP-AGRI: 7 Years of Innovation in Agriculture and Forestry*; European Commission (DG-AGRI): Brussels, Belgium, 2020.
56. Murphy, M.; O'Connell, K. *Systemic Behaviour Change: Irish Farm Deaths and Injuries*; Department of Management & Enterprise Conference Material, Cork Institute of Technology: Cork, Ireland, 2018.
57. Glasscock, D.J.; Rasmussen, K.; Carstensen, O. Psychosocial factors and safety behaviour as predictors of accidental work injuries in farming. *Work Stress Int. J. Work Health Organ.* **2007**, *20*, 173–189. [CrossRef]
58. Teagasc. Farm Safety for Older Farmers. 2021. Available online: https://www.teagasc.ie/news--events/daily/farm-business/farm-safety-for-older-farmers.php (accessed on 1 October 2021).
59. Peters, K.E.; Gupta, S.; Stoller, N.; Mueller, B. Implications of the aging process: Opportunities for prevention in the farming community. *J. Agromedicine* **2008**, *13*, 111–118. [CrossRef] [PubMed]
60. Ponzetti, J. Growing Old in Rural Communities: A Visual Methodology for Studying Place Attachment. *J. Rural Community Psychol.* **2003**, *6*, 1–11.
61. EIP-AGRI. *Agricultural Knowledge and Innovation (AKIS), Systems Stimulating Creativity and Learning*; European Commission (DG-AGRI): Brussels, Belgium, 2018.
62. Agriculture and Horticulture Development Board (AHDB). *Livestock Markets Serving the Industry for 200 Years, A Review of Livestock Markets in England Prepared for the Livestock Auctioneers Association (LAA)*; AHDB Beef and Lamb: Kenilworth, UK, 2017.
63. Nye, C.; Winter, M.; Lobley, M. *More Than a Mart: The Role of UK Livestock Markets in Rural Communities*; Full Report to The Prince's Countryside Fund; The Prince's Countryside Fund: London, UK, 2020.
64. Walsh, K.; O'Shea, E.; Scharf, T. Exploring Community Perceptions of the relationship between age and social exclusion in rural areas. *Qual. Ageing* **2013**, *13*, 16–26.
65. Leonard, B.; Kinsella, A.; O'Donoghue, C.; Farrell, M.; Mahon, M. Policy drivers of farm succession and inheritance. *Land Use Policy* **2017**, *61*, 147–159. [CrossRef]

 *sustainability*

Article

Irish Organics, Innovation and Farm Collaboration: A Pathway to Farm Viability and Generational Renewal

Maura Farrell *, Aisling Murtagh, Louise Weir, Shane Francis Conway, John McDonagh and Marie Mahon

Discipline of Geography, School of Geography, Archaeology and Irish Studies, National University of Ireland, H91 TX33 Galway, Ireland; aisling.murtagh@nuigalway.ie (A.M.); louise.l.weir@nuigalway.ie (L.W.); shane.conway@nuigalway.ie (S.F.C.); john.mcdonagh@nuigalway.ie (J.M.); marie.mahon@nuigalway.ie (M.M.)
* Correspondence: maura.farrell@nuigalway.ie

Abstract: The family farm has been the pillar of rural society for decades, stabilising rural economies and strengthening social and cultural traditions. Nonetheless, family farm numbers across Europe are declining as farmers endeavour to overcome issues of climate change, viability, farm structural change and intergenerational farm succession. Issues around farm viability and a lack of innovative agricultural practices play a key role in succession decisions, preventing older farmers from passing on the farm, and younger farmers from taking up the mantel. A multifunctional farming environment, however, increasingly encourages family farms to embrace diversity and look towards innovative and sustainable practices. Across the European Union, organic farming has always been a strong diversification option, and although, historically, its progress was limited within an Irish context, its popularity is growing. To examine the impact of organic farm diversification on issues facing the Irish farm family, this paper draws on a qualitative case study with a group of Irish organic farmers engaged in the Maximising Organic Production System (MOPS) EIP-AGRI Project. The case study was constructed using a phased approach where each stage shaped the next. This started with a desk-based analysis, then moving on to semi-structured interviews and a focus group, which were then consolidated with a final feedback session. Data gathering occurred in mid to late 2020. Research results reveal the uptake of innovative practices not only improve farm viability, but also encourage the next generation of young farmers to commit to the family farm and consider farming long-term.

Keywords: organics; succession; viability farm collaboration

Citation: Farrell, M.; Murtagh, A.; Weir, L.; Conway, S.F.; McDonagh, J.; Mahon, M. Irish Organics, Innovation and Farm Collaboration: A Pathway to Farm Viability and Generational Renewal. *Sustainability* **2022**, *14*, 93. https://doi.org/10.3390/su14010093

Academic Editor: Giuseppe Todde

Received: 18 November 2021
Accepted: 15 December 2021
Published: 22 December 2021

1. Introduction

In the last four decades, there has been a radical overhaul of the agricultural industry, with a shift from a productivist agricultural regime to a multifunctional agricultural environment [1–3]. Agriculture is considered multifunctional when the functions and services it provides go beyond food production to encompass a wider social, environmental and economic role. This includes, for example, links to local food supply chains, farms that create and preserve cultural landscapes, or preservation of biodiversity, soil and water quality. Multifunctionality can also present itself to different degrees, where highly productivist agriculture is considered to display the least multifunctionality, while agriculture that moves away from the productivist model has strong multifunctionality [2]. Through consistent amendments of the Common Agricultural Policy (CAP), EU policy makers have attempted to deal with a myriad of agricultural issues from environmental concerns to food security. Amongst all this change, the family farm strives to remain relevant, resilient and sustainable [4]. Increasingly composed of a farming populace with a high age profile [5], this ageing community requires an injection of young people into farming by means of efficient and effective intergenerational farm transfer [6]. The perception is often that the older farmers are less competitive in the current marketplace because they are hesitant in their adoption of new practices and innovative agricultural technologies [7,8]. On the other hand, however, the younger generations are looked on as more willing to embrace smart agriculture and alternative farming practices, such as organic agriculture, in striving for a more sustainable, profitable and productive future for farming [9–11]. Consequently, the necessity to get younger farmers engaged in agriculture will not only ensure production efficiency and economic growth of the Agri-food industry, but will be essential to the sustainability of rural society more broadly (ibid).

This paper therefore explores the opportunity presented by organic farming, with a focus on horticulture-based farming systems, in particular, for increased farm viability in the Irish context. In previous research comparing the economic viability of conventional and organic farms in Denmark, Pedersen and Hauge [12] found that 'the profit of conventional farms has decreased, while organic farmers' earnings have increased' (p. 4). Therefore, it is possible to conceive that young farmers or new entrants into farming can enhance farm viability by embracing organic farming, rather than conventional pathways (ibid). The paper also argues that taking advantage of this opportunity is not all straightforward and requires specific supports and understandings. This emergent context highlights the threat to the family farm if transfer to the next generation is not enacted expeditiously. Indeed, Duesberg et al. (2017) argued how farm viability increases the prospect of farm succession and, in turn, enhances the sustainability of the farm. It is in addressing these major interlinked issues of farm viability, reinvigorating agriculture through young and new entrant farmers and addressing the complexity around land transfers and succession that this paper seeks to make its contribution. Drawing insight from the Maximising Organic Production Systems (MOPS) EIP-AGRI (European Innovation Partnership for Agricultural productivity) and Sustainability project, we explore how a change in mind-set can be incubated in the context of a move towards organics. In exploring this, the paper outlines the ways in which organic farming presents an opportunity for improving farm viability through supply chain efficiency and providing 'softer' supports, such as spaces for different types of knowledge (expert and non-expert), sharing and development. The paper also looks more broadly at factors influencing new entrants and succession into organic farming in Ireland, such as specific pathways, including the returning successor, and how wider professional experience and knowledge can benefit the farm in terms of innovation and viability.

2. Literature Review

2.1. Farm Success and Farm Succession

Farm success faces a variety of challenges, including access to land, lack of succession planning, lack of retirement of older farmers and lack of attractiveness of the farming profession [13]. Interlinked issues of concern in this paper are the economic viability of farm livelihoods and how this is very much bound up in aspects of transferability of the farm itself [14]. As such, issues of below-average farm incomes, economically non-viable farms, in addition to farms engaged in pluriactivity are all significant obstacles to new entrants [15–17]. A key starting point for much of this conversation is ensuring that the farm succession process occurs. This, however, is much more complex than dealing with just the actual transfer of ownership. Succession, according to Handl et al. [18] is a multi-faceted, diverse procedure that can occur over a long timescale for an individual or a group. Lobley [13] suggests that farm succession, can also be broken down into 'succession to the farm and succession to the occupation of farming' (p. 839). Consequently, its intricacy has been classified as a multi-stage process, involving the movement from partnership farming to full control, a process that can materialise in a variety of ways [18,19]. The intricacy of succession is further complicated with additional deliberations on defining what a farm successor or, in more recent conversations, what a new entrant is. In fact, an EU, EIP-AGRI Focus Group on New Entrants into Farming found a number of classifications identifying: 'a substantial grey area between the extremes of ex novo new entrants and direct successors to farming businesses' (p. 7). The Focus Group recognised six types of new entrant (diversified new entrant, innovative new entrant, full-time new entrant, part-time new entrant, hobby farmer, hybrid new entrant) and five types of successor (diversifying successor, innovative successor, direct successor, delayed successor, indirect successor), as well as different pathways to these categories. The delayed successor, for example is someone who has worked off-farm and does not make a change to the farm operation [20].

Notwithstanding these issues, another barrier to the success of a farm is the transferability of the farm itself. Larger farms, in particular, are often heavily equipped with expensive machinery and require large amounts of capital in the takeover process, while they are also faced with diminished agricultural revenue. The InPACT [21] project pays particular attention to the French situation and sketches a picture of large farms where farm transfer can lead to alternative farm restructuring, including farm enlargement or even reorientation or land abandonment. To ensure the effective transfer of the farm, a focus on how the farm is transferred is highly significant. Alongside this, matters for new entrants, including the possibility of farm restructuring or diversification to ensure future viability also need consideration.

2.2. Knowledge, Networks and Innovation

A farming context that is facilitative to peer co-learning and knowledge sharing improves farm viability (and, as a consequence, supports the succession of the farm to the next generation). The Access to Land Network [22] paid particular attention to these issues, highlighting the importance of formal, informal and practice-based training for young farmers, including new entrants. Additionally, the Network identified the importance of practice-based learning in consolidating formally learned skills, highlighting key examples within the French context (ibid). The Network also recognised the advantages of the family network and knowledge transfer when compared to the challenges new entrants face in building farming knowledge and skills development. Consequently, they argue that succession within the family farm can provide a space where knowledge, skills and experience can be generated prior to formal succession plans taking place [22,23].

Knowledge requirements and barriers to gaining skills for young farmers depends on a number of factors, including existing education, whether young farmers are new Member State or EU-15, if they are farm owners or the type of rural region they come from [23,24]. More generally, Zondag et al. (ibid) emphasised the importance of knowledge and skills development, particularly in certain areas: 'young farmers need technological skills and skills to develop a farm strategy, as well as entrepreneurial skills—such as marketing, networking, communication and financial skills—to keep their farm viable. They are not always aware that they need all these different kinds of skills. Many farmers are used to managing their farm in a traditional way and do not see the need to change' (p. 70). Focusing on skills development, the Access to Land Network [22] discerns that there is a lack of training, specifically in organic farming, permaculture, or other techniques that can be relevant to new farm entrants. In relation to these 'neo-farmers', Dolci and Perrin [23] note a level of discontent with more conventional, institutional training, and a move towards more informal, alternative sources for skills and knowledge development.

In extending the above discussion, the significance of knowledge creation and innovation has also expanded and is aligned with the need for more informal knowledge systems in farming. However, it is also important to add this is not to discount more scientific and technical innovations, such as through smart farming and ecological innovations, and their relevance to more sustainable and viable future farming. An exploration of innovation theory by Dargan and Shucksmith [25] identified a move away from an emphasis on linear paths, where practitioners apply scientific innovation and novel discoveries towards a diversity of innovative paths, systems and networks. In fact, the practice of 'novelty production' is one of the foundational dimensions of van der Ploeg et al.'s [26] 'rural web' of actors and resources underpinning rural development. In examining the 'rural web', novelty production is defined as a 'capacity, within the region, to continuously improve processes of production, products and patterns of cooperation' (ibid, p. 9). Van de Ploeg et al. (ibid) also emphasised the critical nature of novelty production within a rural sustainability and development context, suggesting they provide 'new insights, practices, artefacts, and/or combinations (of resources, of technological procedures, of different bodies of knowledge) that enable specific constellations (a process of production, a network, the integration of two different activities, etc.) to function better' (p. 9). One key aspect of novelties is their place-based nature, with many based locally, drawing on contextual rather than scientific knowledge, and complementing one another rather than working against each other. In line with this, Dargan and Shucksmith [25] recognized the need and significance of 'bringing together knowledge forms in collective learning processes' (p. 288). Esparcia [27] recognises this and emphasises how knowledge can combine and result in innovation leading to: 'the creation, adoption or adaptation of new knowledge by the actors, combining their initial stock of implicit tacit knowledge with other explicit knowledge (offered or contributed by advisors, consultants, development actors' (p. 288). Tovey [28] and Dargan and Shucksmith [25] also previously highlighted that knowledge can be co-produced and more commonplace; everyday knowledge and learning can be seen as innovation, but adapted or used in an alternative fashion.

Encouraging farmer innovation, the European AgriSpin project also emphasises the importance of collective learning and learning from practice [29]. This philosophy can be captured by the concept of 'vernacular expertise', suggested by Lowe et al. [30] as: 'The expertise people have about the places in which they live and work that is place-based but crucially nourished by outside sources and agents' (p. 36). Vernacular expertise therefore can consist of a diversity of knowledge types made up of both local and extra-local sources. It should not have a hierarchy, with all forms of knowledge (lay or expert, social or scientific) being of equal importance. Additionally, it should be replicable, with the potential to be diffused via multiple pathways, including: peer-to-peer; expert-to-peer; expert-to-practitioner; practitioner-to-practitioner. Lowe et al. [30] highlighted the importance of this type of expertise for rural development, emphasising both its importance and its implications on policy. Similarly, Atterton [31] suggests that a change in policy direction is needed, moving away from central regions as the focus of innovation policy and looking towards nature, potential and needs directly connected to rural innovation, 'recognising that innovations can be small scale and led by an individual with a creative idea to tackle a problem; they need not involve huge R&D expenditure or large numbers of patent registrations' (p. 228). In considering these issues further, this paper draws on an organics case study where a collective space for knowledge sharing demonstrates the possibilities of this approach.

2.3. Organic Farming—An Opportunity for Farm Viability?

The European Commission [32] reported that land under organic production increased by approximately 500,000 hectares annually over a ten-year period, representing a coverage of 11.1 million hectares of European Utilisable Agricultural Area (UAA) [32]. More recently, Eurostat [33] reported the total area under organic farming in the EU increased to cover almost 13.8 million hectares of agricultural land. Although not as popular within an Irish Agri-food context, the organic sector's growth is important in terms of reacting to current marketplace demands and in meeting broader societal expectations. Irish consumer research, for example, has shown an increased inclination towards organic food, in line with a growing trend towards a more health-conscious modern society [34]. Such trends are emulated across the EU, emphasising the opportunities for the enlarged production of organic food products. This trajectory also stresses the health aspects of organic farming, but also the economic, social and environmental benefits of organic systems.

The number of those engaged in Irish organic production increased considerably in recent years, largely due to dedicated policy directives under the Rural Development Programme (RDP) 2014-2020. RDP policies in support of the organic sector have provided €56 million for the Organic Farming Scheme (OFS), while providing area-based payments to registered organic farmers, and an €8 million Organic Capital Investment Scheme, providing grant aid of up to 60% for qualified young organic farmers for investment in structures and equipment. As a result of such policies and increased interest from the farming community and food producers, approximately 72,000 hectares (ha) of Irish farmland is currently certified as organic. This is an increase of nearly 50% since the start of the RDP in 2014 [34]. Additionally, a recent report by the Central Statistics Office (CSO) shows that the area of agricultural land organically farmed in Ireland increased by 257% between 1997 and 2018 [35]. Consequently, Bord Bia, the Irish Food Board, highlight that the organic retail market in Ireland is now worth €162 million, with a further €44 million generated by direct sales [34].

Despite this recent expansion of the Irish organic sector, land under organic production in 2018 still only accounted for 1.4% of the total utilizable agricultural area (UAA), the third lowest percentage among EU Member States (ibid). To address Ireland's organic deficit and respond to the EU's Farm to Fork strategy and its call for 25% of total EU farmland to be utilized for organic farming by 2030, Ireland's national climate and air roadmap for the agriculture sector (Ag Climatise), outlines an ambitious objective of increasing the current area under organic production to 350,000 ha by 2030 [36]. Although the area under organic production has increased, production patterns in Ireland are still not fully in line with market opportunities. In fact, the majority of Ireland's 1700 organic farmers are livestock producers, notwithstanding the fact that organic horticulture, tillage and dairy have been acknowledged by Bord Bia, the Irish Food Board, as having the most significant growth potential in the Irish market [34]. In particular, organic horticulture production is considerably less than what is needed, resulting in almost 70% of organic fruit and vegetables being imported annually to meet market demand. Additionally, there is a supply shortfall of organic cereals and proteins in the Irish market; restricting even further the opportunities for the Irish organic sectors to take advantage of the growth potential that currently exists. Consequently, for the Irish organic food sector to ensure longer-term growth and sustainability, it must be fully cognizant of market desires and buyer demand.

The organic horticulture sector in Ireland is increasingly recognized as one of the organic categories with the highest development potential, with sales of organic fruit and vegetables already representing 34% of the Irish organic market. Westbrook [37] suggests that such figures are reflected in the retail data from other countries, as horticulture is one of the most resilient categories in global organic food sales. Although the ongoing importation of some horticulture goods is essential, given the unpredictable nature of Ireland's climate and the variety of products on offer, Irish farmers are still limited by a lack of capacity to meet the demands for organic horticultural products due to their family-farm, small-scale operations [38]. Farm viability is further compromised with farmers tending to produce similar crops, harvested at related times, which result in surplus produce and wasted goods, which in turn undermines economic performance.

3. Methodology

The research here employed a case study approach, a process that can best be described as 'a methodology, a type of design in qualitative research, an object of study and a product of the inquiry' [39]. The fundamental features of case study research consists of 'a qualitative approach in which the investigator explores abounded system (a case) or multiple bounded systems (cases) over time through detailed, in-depth data collection involving multiple sources of information (e.g., observations, interviews, audio-visual material, and documents and reports) and reports a case description and case-based themes' (ibid.). This is the process which was undertaken in this research, where a case study was utilised to examine issues around young farmers, new entrants, succession, farm viability and organic farming in Ireland. The case study drew on the experience of key stakeholders involved in Irish organic farming, including farmers, extension advisors, Department of Agriculture officials, policy makers and horticulturists. A core focus of the case study however, revolved around an Irish European Innovation Partnership in Agriculture (EIP-AGRI) organics project, titled the Maximising Organic Production System (MOPS).

The case study was carried out from the beginning of June 2020 to the end of October 2020. A desk-based study was initially carried out (grey literature, online evidence, policy documents and Central Statistics Data) to collect background information on the Irish organic industry as well as the MOPS project. The case study employed a three-pronged approach, initially carrying out twenty-two in-depth, semi-structured interviews, eleven of which were with individuals engaged in the MOPS project, and the remaining eleven interviewees made up of additional organic farmers, horticulturists, project administration, policy makers and extension advisors. All interviewees were selected based on the desk-based analysis and also by using a snowball method. The second element of the methodology involved a focus group, consisting again of key personnel engaged in both the MOPS project and organic farming, and this was used to gather further information for data triangulation. Finally, the third element involved a findings feedback session with similar personnel. In addition to disseminating some initial research results, the final feedback session also allowed a further exploration of issues that were not fully examined within the first two phases of the research methodology. Interview questions for all three-research elements were prepared based on the key research agenda related to how organic farming, and the MOPs project in particular, contribute to generational renewal at a farm level. All interviews were recorded and fully transcribed, and then a coding system was devised via Nvivo, where the analysis was carried out using a thematic analysis approach. Due to COVID-19 restrictions in Ireland, all interviews, the focus group and feedback session were held online via Zoom. In all three cases, a gender balance was considered, resulting in an equal amount of male and female participants, where possible.

Maximising Organic Production System (MOPS)

To help situate the MOPS project in its geographic, structural and formative context, we next outline its origins and the locations of the farms involved, as well as its broad structural and operational characteristics. One clear characteristic of the group is its diversity—of the markets supplied, the geographic location and the sizes of the farms. This shows how collaborative projects can work with dispersed and varied organic horticulture farms.

The origins of the MOPS project are rooted in an agronomy group coming together originally to seek advice from a well-known agronomist specialising in organic farming in Ireland. However, this process also brought wider shared concerns to light, such as how working in partnership could improve their farm's economic sustainability, as well as having spin-off benefits to improve farmers' work-life balance. A call from the Department of Agricultural Food and Marine (DAFM) for EIP-AGRI Operational Groups provided a fitting vehicle to formalise the group and to support it with funding. A group of organic farmers, the Irish Organic Association (IOA), researchers and agronomists came together and formed the MOPS Operational Group that successfully obtained €597,416 in funding from the DAFM to run a three-year project. MOPS became one of Ireland's first EIP-AGRI Operational Groups in 2018. MOPS is one of the first EIP-AGRI organic projects in Ireland engaged in creating a short supply chain for their farm produce.

More specific objectives of MOPS included improving economic sustainability and farm viability. A key action to support this was the development and application of cropping systems tailored to each farm focused on achieving greater efficiency within production, such as more continuity year-round and production that was closely tailored to market demand. Improving economic sustainability also looked beyond the farm gate. Actions worked to improve the knowledge of future market demands, as well as focusing on improving short supply chain efficiencies. The project objectives also focused on improving the environmental sustainability of the farms through reducing the use of imported nutrients and increasing the use of green cover crops.

The farms themselves and their geography is quite diverse. The geography of the MOPS group is dispersed, crossing a number of NUTS 3 regions in the east, west and south of Ireland. Specifically, this represents seven counties (Kilkenny, Cork, Galway, Laois, Wicklow, Kildare and Wexford). The farms are all certified organic, operating on either leased or inherited land. The 11 MOPS farms are owner operated, however, a number also lease land. They are a mix of relatively recent (last 5 years) to longer established (last 20 years) organic growers. They also vary in size from one to three-hundred-hectare farms. Collectively this group of farmers utilises a range of marketing channels: direct sales (online and farmers markets), wholesale markets, speciality shops, restaurants, private procurement outlets, and supermarket/retail multiples. The farms achieve a year-round supply of crops, but also import organic produce to supplement their farm produce (Westbrook, 2020). Beyond their horticultural activities, many engage in multifunctional farming, including training organic growers via an apprenticeship programme.

4. Findings and Discussion

4.1. Enabling Realisation of the Organic Farming Opportunity

4.1.1. Improving Farm Viability through Supply Chain Efficiency: A Collective Approach

A core focus of the MOPS project is to optimise organic horticulture. This encompasses the creation of a collaborative cropping system responding to growing retail demand, and improving the continuity of short supply chains for the national market. This involves exploring each farmers' cropping system and identifying what is most profitable based on market demand. However, crucially, this is also based on crops suited to the farm, the farmer and their skillset. For example, a pioneer of the MOPS project states: "More important than profitability that they actually have a demand for those crops but also that it suits what they have on their farm, it suits their skillset, it suits their employment and what staff they have, their machinery and all the various other things that impacts on the capacity of that farm" (Interviewee, 1).

MOPS facilitates the farms involved to explore and move towards more profitable crops, as opposed to over-producing certain crops that lead to waste and financial losses. Also, at the farm level, it aims to deliver a tailor-made cropping system that is flexible to market demands and increases profit on farms. It drives farms to reflect on their economics, which ironically can be overshadowed due to time limitations: "One of the aims of MOPS (being) to go through the figures and actually see what you're doing that is not so profitable and what's more profitable. Because sometimes it's quite difficult when you're actually busy with it to actually differentiate which crop is really making the money and which is not you know" (Interviewee, 4). Market demand is considered at-scale, which also facilitates a range of different farm sizes to be part of the project. Some of the MOPS group concentrate on local, smaller markets while others concentrate on larger markets. For example, one farmer saw himself as; "a commercial grower of the MOPS group" while also commenting: "then there's other really good independent growers that do a lot of box schemes" (Interviewee, 19). Alongside this focus on improving economic profitability, MOPS is also concerned with making farms more sustainable through reducing nutrient import dependency, which also supports this aim.

Beyond the primary goals of improving profitability through short supply chains and enhancing training, many interviewees were quick to point out the project's added value in supporting farm viability. For example, it enabled farmers to explore more efficient organic farming practices and provided a space for connection with other farmers to enhance their current practices. One farmer felt that MOPS also had a wider value in the sector's viability: "MOPS has done huge work for the organic horticulture sector in Ireland, just in joining the dots and making sure this farmer is growing that and this farmer needs it or this market needs it here. There is nothing more disheartening in doing something and wasting a crop or not being able to sell it all" (Interviewee 8). Additionally, at the individual farm level, enhancing farm viability was a key aspect. For example, innovative thinking drove efforts to increase income and long-term viability for one farmer: "I launched a veg box scheme in 2019. Literally just one night decided to set up an Instagram page, a Facebook page with a logo on it and just put it out there. And I would say within a week we were booked solid" (Interviewee, 14). The case study underscores the relevance of Pedersen and Hauge (2016) findings that emphasise the financial viability of organic farms over conventional farms, with increased earnings due to novel and innovative farm practices.

4.1.2. Networking and Knowledge Sharing

Knowledge development is a key benefit of this project. MOPS gave the group the opportunity to work with suitably qualified people over its three-year span. The value of this additional support was noted by many interviewees, with one suggesting: "The MOPS project and advisors have been really good for evaluating which crops work best for us and which don't" (Interviewee, 9). While another suggested: "Well the very first simplest thing is because we've had to submit records religiously, records of you know what we sow, when we sow it, how long it took us to sow it, when we harvested it, how long it took us to harvest, how much we made from it, how much is left, how much was spoiled you know all these details. Initially it was a nightmare for us because it is just all this work like capturing everything. But with time we realised it actually was really helping us. We were a bit less stressed. You know you could actually check. You know you could go back a few months. You could see exactly what day was what. Rather than holding everything in your head" (Interviewee, 10). This also shows how knowledge and skills acquisition to improve farm viability can be related to more straightforward management and record-keeping issues. These are issues that can be quickly addressed with the right training methods.

Another way MOPS supported knowledge development was through peer-to-peer learning. The organic farmers interviewed highly valued the MOPS collaborative model also because it provided them with an ideal and continuous forum to engage with each other. One farmer in particular stated that; "I suppose the good thing about MOPS as much as anything else is that we have constant advice and we have also the kind of teamwork. I won't say its teamwork in that we're all exactly on the same hymn sheet but you know the way ... You've somebody to consult and talk to about things you know" (Interview 4). The collaboration and networking among the group also facilitated technical learning based on different on-farm experiences. Another interviewee stated: "A WhatsApp group came out of it and then there'd be meetings ... you would just pick up tips and run things by people and you'd help others as well like no we did that variety and these are the issues we had with that and you know they don't grow well in this kind of soil. You know it's just so nice you don't have to invent the wheel. Like there's this wealth of knowledge and expertise and you can all help each other" (Interview 9). Knowledge development also went beyond the technical. There was also a wider change in attitude. One respondent being stated: "No it wasn't really upskilling so much as just changing our attitude really to what's important actually this is important" (Interviewee, 9).

Another strength of MOPS is the space provided for intergenerational knowledge transfer and sharing across farming generations. Within the MOPS group are established older farmers that hold an invaluable store of tacit and who lay knowledge developed over years of hand-on working in the organic sector. The younger generation have not had the time or experience to develop this knowledge. For example, this interview illustrates: "There's a generation of them there all in their sixties ... years and years of experience ... It's vital you know that transfer of knowledge ... I've said it at multiple meetings ... The knowledge, the boots on the ground of going out to a field and looking at a crop and saying that's what's wrong with this crop... I will say my expertise would be in carrots because I've grown up with them. I've seen every single different breed and disease and condition. And I could walk out into a field and I could say that's what's wrong with those or X Y and Z just through years of experience. I couldn't do the same for broccoli or cauliflower. Now I'm learning" (Interviewee, 19). Laband and Lentz [40] highlight he importance of making such 'soil-specific' human capital more easily transferable, communicated or learnable. MOPS provides a space for the nurturing of the younger farmers' enthusiasm and ambition, guided by the senior generation. There is: "a level of communication with new entrants in a very practical and worthwhile way" (Interview 6).

4.1.3. The Value of Innovation

Undervaluing the on-site research and innovative ideas of farmers is an issue impacting innovation levels in farming. For example, Kummer et al. [41] argued that the innovations of organic farmers can be ignored, despite their ability to lead to strategic farm changes and hence their significance. The ethos and approach of MOPS farmers and those in organic farming more widely is innovative and focuses on the transferability of ideas.

Most interviewees, particularly within MOPS, but also the other organic farmers, discussed how they put innovative ideas into practice. They held a strong awareness of the necessity of innovation to producing high quality organic produce and effectively using local supply distribution channels. New ideas and innovations that could be classed as social, technological or product innovations were developed, enhancing family farm viability and work-life balance.

In addition, an important pattern was that more traditional scientific innovation was not always central. What emerges more strongly is more everyday innovation and new ways of doing things. For example: "I suppose the innovation doesn't all have to be highly scientific stuff. I'd say the innovation wasn't what we expected. The innovation has actually come about by just having good record keeping and maybe considering using a database or kind of a gatekeeper is actually quite complicated for a lot of the group, but you know to use some sort of system in order to keep track of things" (Interviewee, 1). New ways of doing things while requiring some adaptation and change, bring important efficiencies for example: "It means everything in the packaging houses are all packed on tablets, touch screens and there's no paper anymore. All the paper has been eliminated—everything is automated from the order to the payments to the packing to the delivery. It's all on apps and that was a big project for us you know but it meant like thousands and thousands of print-outs and double checking just disappeared literally overnight". (Interviewee, 6). Everyday innovation and new ways of doing things can also mean simple changes, but which have significant impacts, such as relating to crop waste for example: "Its things like WhatsApp. But actually having a WhatsApp group that is telling you that I have X amount of parsnips or whatever available and so trying to reduce your waste. The innovation is more in the approach than it is in an actual piece of technology" (Interviewee, 1).

Further to this, the presence of strong innovation in a farming sector may also be a factor in attracting new entrants into farming. This emerged in the focus group: "It's about this innovativeness of organic farmers and even the collective innovation and the different synergies and how that you know can it play a role in enticing the younger farmer. Maybe the innovative practices in organics, can entice younger farmers into farming but of course obviously into organic farming" (Focus Group Member).

4.2. Linking Succession Patterns and Organic Farming—Aspects of the Returning Successor

Given the issue of farm succession in farming more widely is a key challenge for the sector, it is worthwhile exploring if aspects of organic farming and the MOPS project may provide ways to improve succession. However, the pattern of intergenerational succession and eventual farm inheritance emerged here as the main route of entry into organic farming. Traditional patterns of intergenerational family farm transfer appear to persist in organic farming. One MOPS farmer suggested: "I basically went into full-time farming working with my father and I would have worked alongside him for a number of years and then when I was in my mid-twenties I started renting land from my father at first" (Interviewee, 5). Another organic farmer stated: "Well I grew up on a farm, so I've been on the farm all my life and interested in the farm all my life. I've never actually done anything else... I went from school on to the farm. My grandfather was a farmer so it's in the family and it would have been I suppose just the only thing I really wanted to do. I suppose getting into the farming then I was kind of happy just working on the farm. I was working for my dad and he passed away in 2003 so I took over the farm then and I started farming it" (Interviewee, 12). Another MOPS farmer engaged in organic vegetable growing stated: "My father and mother they were elderly ... and going to give up farming so I just started farming four acres of organic veg here. And then after the first year he just made about three quarters of the farm over to me and a quarter to my other brother. So that's kind of where it all started from" (Interviewee, 20).

The reasons for this traditional pattern of intergenerational family farm transfer appear to be linked to the presence of, as described by Conway et al. [6], a deeply ingrained 'rural ideology' where farm succession within the family is prioritised. From this research, most of the organic farms showed traditional patterns of farm succession and inheritance. However, this research also revealed another specific pattern in how succession occurred. Successors can also leave the farm and return after a period of travel, study and or work off-farm and outside of the agricultural sector. Errington [42] call this pattern a 'professional detour'. While this is not a departure from traditional patterns, it is one that appears to benefit the farm business, so could be an important route to support for both increased succession, as well as greater farm viability.

Interview data show that individuals with diverse careers that young farmers, outside of farming, pursued before returning, such as in the pharmaceutical, construction and hospitality industry. It is not just the professional experience gained that is of benefit, but also the experience increased ambition and impacted a vision emerging to convert to an organic farming system. Also supporting this was greater courage and confidence to make a significant change on the family farm to support its long-term sustainability and viability. For example, one MOPS farmer stated: "I got a degree in applied chemistry in Galway. Then went to do a PhD in Cambridge in the U.K and then went into the pharmaceutical and the biotech industry in England for eleven years. So you know I'd be very aware of chemistry, chemicals, biochemistry, the background to you know how chemicals work in the environment and potential pitfalls of using them etc." (Interviewee, 6). Another organic farmer stated that, "I worked around the world. I worked in Germany. I worked in Spain. I worked in Australia ... So that's what I was at before I came home and took over the farm ... So, I had seen an awful lot of diversity in farming around the world and little small farmers up on hilltops in India and in Nepal and how they were making a living off a very small part of the land. So that put me on a journey towards organic when I see how they were viable" (Interviewee, 11). Similarly, another organic farmer stated that: "I did four years in the bank after Edinburgh and then there was more and more helping needed at home so I needed to kind of be at home more so I looked at retraining. Went up to Donegal and did a FAS course up there for a year and a half in stone masonry. And at least then I had a trade that I could be self-employed with and kind of fit it in around the farm ... all these things are important rather than just having a very small realm of experience doing what your father did kind of thing" (Interviewee, 13). Further still, a MOPS member farmer outlined: "I did social science in Dublin and I lived there for a number of years and worked in that area ... I worked in that kind of industry in rehabilitation and all that kind of stuff. And then I decided to move back home and took on part of the family farm and I started basically to grow organic vegetables" (Interview, 21).

Favoured as a more sustainable production system, research has found new entrants are more likely to pursue organic farming [9,43–46]. This research also shows the inclination of the returning successor towards organic farming. Based on an analysis of the socio-economic impact of organic and non-organic farmers in England, Lobley et al. [9,47] found that significant numbers entered organic farming as a completely new career, who also often had urban origins, arguing they potentially represent a new agricultural paradigm. This research shows this is not a generalisable pattern. The findings follow Rigby et al. [43] who argued that, on average, organic farmers enter farming later in life. More broadly, the findings echo research that highlights the characteristics of new entrants, as opposed to successors. Zagata et al. [48] argued that new entrants to agriculture, of any age, are potential innovators. This highlights the wider value of new entrants into organic farming, regardless of whether they are from a farming or non-farming background. Sutherland et al. [46] also argued the positive effect of new entrants to organic farming because they can be more entrepreneurial, business orientated and proficient in setting up new market opportunities.

More broadly, experience working outside of farming brings wider benefits. Interview data show how this experience shaped the farmers' skillsets and ability to overcome key challenges facing agriculture. Farmers appeared to have strong capacities to manage and operate a profitable organic farming enterprise, regardless of its size and scale. Small-scale organic farmers appeared well-equipped to overcome challenges also faced by conventional farmers. This included access to land and challenges gaining adequate capital access to compete in scale-driven markets. These farmers became involved in niche markets, finding novel ways to reach their consumers. For example, this could include box schemes, farm shops, farmers' markets and on-farm processing facilities. For example, one MOPS farmer stated: "We don't have loads of acres and we don't have access to cheap labour. So my feeling was that we needed to do something a little bit different and a little bit more high value at the end and something niche and I suppose that's where I got the interest in organic farming" (Interviewee, 5). More broadly in Ireland, farmers' markets appear to be an effective channel for selling organic produce, particularly because the customer base tends to share the ethos held by organic farmers [49]. When compared to specialised conventional operations, the data also show many of the organic farmers grow a diverse range of horticultural produce, which also needs to be supported by a wide skillset. This is an important approach, helping to spread risk and exposure to external forces, such as harsh weather which is unfavourable to horticulture or market fluctuations. Furthermore, interview participants also felt Ireland's green international reputation strengthens their platform on which to develop their organic enterprises. These activities combine to strongly support the farm income of interview participants.

This section may be divided by subheadings. It should provide a concise and precise description of the experimental results, their interpretation, as well as the experimental conclusions that can be drawn.

5. Conclusions

This paper identifies the positivity emerging and the opportunities presented by organic farming as a route to help increase farm viability in the Irish context. What has also become apparent is that there are opportunities, not only in the development trajectory of existing organic producers, but, importantly, that organic farming can act as a catalyst in attracting new entrants to the agricultural sector. This is a particularly important aspect as it will most likely become a vital contributor in ensuring that the ambitious and challenging growth projections for the industry which are set out in the European Union's Farm to Fork strategy are met. The MOPS project presents a model that can support the greater economic and environmental sustainability of organic horticulture in Ireland. The collaborative production it supports on farms assists horticulture producers to tailor their production to market demands. The focus on the increasing use of green cover crops and minimising external nutrient inputs supports a greater environmental sustainability. The social sustainability of the wider farming environment in Ireland is also supported by organic farming as a potential catalyst for greater levels of succession and the attraction of new entrants into farming.

The case study in this research also demonstrates how dealing with issues related to farm viability at the collective level is effective at improving viability, and has spin-off knowledge sharing and innovation benefits that also support this aim. The transferability of the case more widely within organic horticulture would likely have benefits, as the sector has market opportunities. Access to land also emerged here as an issue, alongside issues specific to succession, the need for tailored supports and dealing with the perception of organic farming.

Another important finding is the pathway into organic farming of the returning successor. This has relevance for policy. Targeted support that incentivises the returning successor could attract those back into farming who have left to pursue education and employment elsewhere. This is potentially a broader way forward for policy, which identifies different types of successors and new entrants, and targets supports specifically to their needs. However, there is also a need for further research understanding the specifics of how succession effectively occurs, and how new entrants get into farming. Research exploring the categorisations (the six types of new entrant and five types of successor) identified by the EIP-AGRI [20] Focus Group on New Entrants into farming is perhaps a potential starting point towards understanding the needs of successors and new entrants more specifically

The MOPS case study also demonstrates the potential of group cooperation to support farm viability and succession. The capacity for collective groups to support smaller, emerging sub-sectors of farming to meet their knowledge needs (e.g., speciality beef producers, hemp growers) alongside supporting supply chain innovation emerges as a potentially important focus of supports worth further examination.

Author Contributions: Conceptualisation, M.F., A.M., L.W., S.F.C.; Data curation, M.F., L.W., A.M., S.F.C.; Methodology, M.F., A.M., L.W.; Writing original draft, M.F., A.M.; Review and Editing, M.F., A.M., L.W., S.F.C., J.M. and M.M. All authors have read and agreed to the published version of the manuscript.

Funding: This research was carried out as part of the RURALIZATION project. It received funding from the European Union's Horizon 2020 research and innovation programme under grant agreement No 817642. The opinions expressed in this document reflect only the author's view and in no way reflect the European Commission's opinions. The European Commission is not responsible for any use that may be made of the information it contains.

Institutional Review Board Statement: RURALIZATION has received approval from the National University of Ireland Research Ethics Committee (application number: 20-Apr-08).

Informed Consent Statement: Informed consent was obtained from all subjects involved in the study.

Data Availability Statement: The interview transcript and focus group data on which this paper is based are not publicly available due to ethical consideration and requirements around the protection of personal data based on the General Data Protection Regulation of the EU.

Acknowledgments: The authors wish to acknowledge the support given by the Maximising Organic Production System (MOPS) EIP-AGRI Project team when carrying out this research.

Conflicts of Interest: The authors declare no conflict of interest.

References

1. Wilson, G.A. The spatiality of multifunctional agriculture: A human geography perspective. *Geoforum* **2009**, *40*, 269–280. [CrossRef]
2. Wilson, G.A.; Burton, R.J. 'Neo-productivist'agriculture: Spatio-temporal versus structuralist perspectives. *J. Rural Stud.* **2015**, *38*, 52–64. [CrossRef]
3. Bianchi, R. Multifunctionality in agriculture. New productive paths in the primary sector. *J. Agric. Life Sci.* **2018**, *5*, 1–7. [CrossRef]
4. Darnhofer, I.; Lamine, C.; Strauss, A.; Navarrete, M. The resilience of family farms: Towards a relational approach. *J. Rural Stud.* **2016**, *44*, 111–122. [CrossRef]
5. Duesberg, S.; Bogue, P.; Renwick, A. Retirement farming or sustainable growth–land transfer choices for farmers without a successor. *Land Use Policy* **2017**, *61*, 526–535. [CrossRef]
6. Conway, S.F.; McDonagh, J.; Farrell, M.; Kinsella, A. Human dynamics and the intergenerational farm transfer process in later life: A roadmap for future generational renewal in agriculture policy. *Int. J. Agric. Manag.* **2019**, *8*, 22–30.
7. Ingram, J.; Kirwan, J. Matching new entrants and retiring farmers through farm joint ventures: Insights from the fresh start initiative in Cornwall, UK. *Land Use Policy* **2011**, *28*, 917–927. [CrossRef]
8. Conway, S.F.; McDonagh, J.; Farrell, M.; Kinsella, A. Mobilising Land Mobility in the European Union: An Under-Researched Phenomenon. *Int. J. Agric. Manag.* **2020**, *9*, 7–11.
9. Lobley, M.; Butler, A.; Reed, M. The contribution of organic farming to rural development: An exploration of the socio-economic linkages of organic and non-organic farms in England. *Land Use Policy* **2009**, *26*, 723–735. [CrossRef]
10. CEJA; DeLaval. European Young Farmers—Building a Sustainable Sector 2017. Available online: https://wordpress.ceja.eu/wp-content/uploads/2019/09/CEJA-Delaval-Survey.pdf (accessed on 10 March 2021).
11. Ferreira, S.; Oliveira, F.; Silva, F.G.; Teixeira, M.; Gonçalves, M.; Eugénio, R.; Damásio, H.; Gonçalves, J.M. Assessment of Factors Constraining Organic Farming Expansion in Lis Valley, Portugal. *Agri. Eng.* **2020**, *2*, 8. [CrossRef]
12. Pedersen, M.Z.; Hauge, L.M. Farmers' Attitude Towards Succession and Unconventional Tenures—Providing relevant on unconventional tenures for new entrants. The Digital Project Library. Aalborg University. 2016, pp. 1–146. Available online: https://projekter.aau.dk/projekter/da/studentthesis/farmers-attitude-towards-succession-and-unconventional-tenures(e5918e37-2613-45b0-90b4-93866d23aac4).html (accessed on 20 October 2021).

13. Lobley, M. Succession in the Family Farm Business. *J. Farm Manag.* **2010**, *13*, 839–851.
14. Conway, S.F.; McDonagh, J.; Farrell, M.; Kinsella, A. Uncovering obstacles: The exercise of symbolic power in the complex arena of intergenerational family farm transfer. *J. Rural Stud.* **2017**, *54*, 60–75. [CrossRef]
15. Conway, S.F.; McDonagh, J.; Farrell, M.; Kinsella, A. Till death do us part: Exploring the Irish farmer-farm relationship in later life through the lens of 'Insideness'. *Int. J. Agric. Manag.* **2018**, *7*, 1–13.
16. Regidor, J.G. EU Measures to Encourage and Support New Entrants. DG Internal Policies Policy Department B Structural and Cohesion Policies. Available online: https://www.europarl.europa.eu/RegData/etudes/note/join/2012/495830/IPOL-AGRI_NT%282012%29495830_EN.pdf (accessed on 2 February 2020).
17. Van Vliet, J.A.; Schut, A.G.; Reidsma, P.; Descheemaeker, K.; Slingerland, M.; van de Ven, G.; Giller, K.E. De-mystifying family farming: Features, diversity and trends across the globe. *Glob. Food Sec.* **2015**, *5*, 11–18. [CrossRef]
18. Handl, B.; van Boxtel, M.; Hagenhofer, K. The Process of Farm Transfer. In *Farm. Succession: Tools and Methods to Promote a Successful Farm. Succession. Examples from France, Belgium, Austria and the Netherlands*; van Boxtel, M., Hagenhofer, K., Handl, B., Eds.; European Commission: Brussels, Belgium, 2016; pp. 6–8. Available online: https://ec.europa.eu/programmes/erasmus-plus/projects/eplus-project-details/#project/2014-1-FR01-KA204-002570 (accessed on 12 November 2019).
19. Chiswell, H. From Generation to Generation: Changing Dimensions of Intergenerational Farm Transfer. *Sociol. Ruralis* **2018**, *58*, 104–125. [CrossRef]
20. EIP-AGRI. New Entrants into Farming: Lessons to Foster Innovation and Entrepreneurship. EIP-AGRI Focus Group Final Report. 2016. Available online: https://ec.europa.eu/eip/agriculture/sites/agri-eip/files/eipagri_fg_new_entrants_final_report_2016_en.pdf (accessed on 9 January 2020).
21. InPACT. Des idées pour transmettre Si on restructurait les fermes? 2019. Available online: https://terredeliens.org/trois-nouvelles-publications-sur.html (accessed on 28 April 2020).
22. Access to Land Network. Europe's New Farmers: Innovative Ways to Enter Farming and Access Land. 2018. Available online: https://www.accesstoland.eu/Access-to-land-for-new-entrants (accessed on 27 January 2020).
23. Dolci, P.; Perrin, C. Neo-Farmers: Drivers of Farming Systems Innovation and of the Transition to Agro-Ecology? The Case of Alentejo (Portugal). European IFSA Symposium 2018. Available online: https://prodinra.inra.fr/?locale=fr#!ConsultNotice:435462 (accessed on 27 January 2020).
24. Zondag, M.J.; Koppert, S.; de Lauwere, C.; Sloot, P.; Pauer, A. Needs of Young Farmers: Report I of the Pilot Project: Exchange programmes for Young Farmers. Report produced for the European Commission. 2015. Available online: https://ec.europa.eu/info/food-farming-fisheries/key-policies/common-agricultural-policy/evaluation-policy-measures/farmers-and-farming/pilot-project-exchange-programmes-young-farmers_en (accessed on 12 August 2019).
25. Dargan, L.; Shucksmith, M. Leader and Innovation. *Sociol. Ruralis* **2008**, *48*, 274–291. [CrossRef]
26. van der Ploeg, J.D.; van Broekhuizen, R.; Brunori, G.; Sonnino, R.; Knickel, K.; Tisenkopfs, T.; Oostindie, H. Towards a framework for understanding regional rural development. In *Unfolding Webs: The Dynamics of Regional Rural Development*; van der Ploeg, J.D., Marsden, T., Eds.; Van Gorcum: Assen, The Netherlands, 2008; pp. 1–28.
27. Esparcia, J. Innovation and networks in rural areas. An analysis from European innovative projects. *J. Rural Stud.* **2014**, *34*, 1–14. [CrossRef]
28. Tovey, H. Introduction: Rural Sustainable Development in the Knowledge Society Era. *Sociol. Ruralis* **2008**, *48*, 185–199. [CrossRef]
29. Wielinga, E.; Paree, P. AgriSpin Deliverable 2.4—The Cross Visit Method: An Improved Methodologic Approach. 2016. Available online: https://agrispin.eu/wp-content/uploads/2016/11/Cross-Visits_Improved-Methodology-1.pdf (accessed on 7 October 2019).
30. Lowe, P.; Phillipson, J.; Protor, A.; Gkartzios, M. Expertise in rural development: A conceptual and empirical analysis. *World Dev.* **2019**, *116*, 28–37. [CrossRef]
31. Atterton, J. Invigorating the New Rural Economy: Entrepreneurship and Innovation. In *Routledge International Handbook of Rural Studies*; Shucksmith, M., Brown, D., Eds.; Routledge: London, UK, 2016; pp. 217–235.
32. EC Action Plan for the Future of Organic Production in the European Union. European Commission. 2014. Available online: https://ec.europa.eu/agriculture/organic/sites/orgfarming/files/docs/body/act_en.pdf (accessed on 5 August 2015).
33. Eurostat. Organic farming Statistics 2019. Statistics Explained. The Statistical Office of the European Union. Available online: https://ec.europa.eu/eurostat/statistics-explained/index.php?title=Organic_farming_statistics (accessed on 14 June 2021).
34. DAFM (Department of Agriculture, Food and the Marine). 2019; Review of Organic Food Sector and Strategy for Its Development 2019–2025.
35. CSO (Central Statistics Office). Environmental Indicators Ireland 2020. Available online: https://www.cso.ie/en/releasesandpublications/ep/p-eii/environmentalindicatorsireland2020/ (accessed on 12 August 2021).
36. DAFM (Department of Agriculture, Food and the Marine). *Ag Climatise, A Roadmap towards Climate Neutrality*; Department of Agriculture, Food and the Marine: Dublin, Ireland, 2020.
37. Westbrook, G. Maximising Organic Production Systems (MOPS) Meeting the Growing Demand for Organic Fruit and Vegetables in Ireland Multi-Actor Perspectives, 2020: EIP-AGRI Guest Blogs—Rural Development Programme (RDP) 2014–2020, National Rural Network, Department of Agriculture, Food and the Marine. Available online: https://www.nationalruralnetwork.ie/eip-agri-blog/maximising-organic-production-systems-mops/ (accessed on 28 April 2020).
38. Moroney, A.; O'Reilly, S.; O'Shaughnessy, M. Taking the leap and sustaining the journey: Diversification on the Irish family farm. *J. Agric. Food Syst. Community Dev.* **2016**, *6*, 103–123. [CrossRef]

39. Creswell, J.W.; Hanson, W.E.; Plano Clark, V.L.; Morales, A. Qualitative Research Designs: Selection and Implementation. *Couns. Psychol.* **2007**, *35*, 236–264. [CrossRef]
40. Laband, D.N.; Lentz, B.F. Occupational Inheritance in Agriculture. *Am. J. Agric. Econ.* **1983**, *36*, 311–314. [CrossRef]
41. Kummer, S.; Leitgeb, F.; Vogl, C.R. Farmers' own research: Organic farmers' experiments in Austria and implications for agricultural innovation systems. *Sustain. Agric. Res.* **2017**, *6*, 526. [CrossRef]
42. Errington, A. Handing over the reins: A comparative study of intergenerational farm transfers in England, France and Canada', 2002. In Proceedings of the Xth EAAE Congress, Exploring Diversity in the European Agri-Food System', Zaragoza, Spain, 28–31 August 2002.
43. Rigby, D.; Young, T.; Burton, M. The development of and prospects for organic farming in the UK. *Food Policy* **2001**, *26*, 599–613. [CrossRef]
44. Padel, S. Conversion to Organic Farming: A Typical Example of the Diffusion of an Innovation? *Sociol. Ruralis* **2001**, *41*, 40–61. [CrossRef]
45. Mayen, C.; Balagtas, J.; Alexander, C. Technology Adoption and Technical Efficiency: Organic and Conventional Dairy Farms in the United States. *Am. J. Agric. Econ.* **2010**, *92*, 181–195. [CrossRef]
46. Sutherland, L.A.; Zagata, L.; Wilson, G.A. Conclusions. In *Transition Pathways Towards Sustainability in Agriculture: Case Studies from Europe*; Sutherland, L.-A., Darnhofer, I., Wilson, G.A., Zagata, L., Eds.; CABI: Wallingford, CT, USA, 2015; pp. 205–220. Available online: https://bit.ly/3Ejk2gf (accessed on 25 September 2021).
47. Lobley, M.; Reed, M.; Butler, A.; Courtney, P.; Warren, M. *The Impact of Organic Farming on the Rural Economy in England Final Report to DEFRA CRR Research Report No. 11*; University of Exeter: Exeter, UK, 2005.
48. Zagata, L.; Hrabák, J.; Lošťák, M.; Bavorová, M.; Ratinger, T.; Sutherland, L.; McKee, A. *Research for AGRI Committee—Young Farmers Policy Implementation after the 2013 CAP Reform*; European Parliament, Policy Department for Structural and Cohesion Policies: Brussels, Belgium, 2017; Available online: http://www.europarl.europa.eu/RegData/etudes/STUD/2017/602006/IPOL_STU(2017)602006_EN.pdf (accessed on 21 June 2019).
49. Bord Bia. Guide to Selling Through Farmers' Markets, Farm Shops and Box Schemes in Ireland. 2007. Available online: https://www.bordbia.ie/globalassets/bordbia.ie/lifestyle/information/farmer-markets/guide-to-selling-through-farmers-markets.pdf (accessed on 20 March 2021).

 MDPI

Article

Designation, Incentivisation and Farmer Participation—Exploring Options for Sustainable Rural Landscapes

John McDonagh

Department of Geography, School of Geography, Archaeology & Irish Studies,
National University of Ireland (NUIG), H91 TK33 Galway, Ireland; john.mcdonagh@nuigalway.ie

Abstract: This paper explores how policies, management practices and farmer participation can help shape resilient and sustainable rural environments. The key elements of the paper draw together policy initiatives on land designation for conservation, action-based and results-based agri-environment programmes, and locally led and inclusive partnership models. In doing this, the paper explores ways to address environmental decline, while also allowing farmers to farm and for management practices to be developed that farmers can readily endorse. The paper draws from empirical evidence gathered from two case studies in West Ireland. These studies include in-depth interviews, consultations with key stakeholders and an exploration of policy and other documentation associated with the management of rural landscapes. What emerges from the discussion and the field evidence is that, while there can be discontent, even disillusionment with some practices, there are models of great promise evolving. In particular, the research identifies the importance of enabling a space in which a farmer's knowledge and expertise have due prominence and where they are afforded recognised input in the schemes being developed and promoted. The conclusion of the paper suggests that, while impacts vary, it is clear that combining forces from top-down and bottom-up, allied to locally led decision-making input, provides the Special Areas of Conservation combination whereby landscapes can be both farmed and protected.

Keywords: Special Areas of Conservation; designation; results-based payments; farmer participation

Citation: McDonagh, J. Designation, Incentivisation and Farmer Participation—Exploring Options for Sustainable Rural Landscapes. *Sustainability* **2022**, *14*, 5569. https://doi.org/10.3390/su14095569

Academic Editor: Gideon Baffoe

Received: 29 March 2022
Accepted: 29 April 2022
Published: 5 May 2022

1. Introduction

The rural is at a time when opportunities are plentiful and numerous threats and challenges are ever-present. The rural is talked of as a panacea for addressing major challenges of climate change and food and energy security, while also being looked on as the main contributor to climate change, threats to biodiversity and broader ecosystem destruction. While both carry merit, the rural, for all its missteps and mistakes, is central to any possibility of a sustainable future. The key lies in the ambition and will to make decisions that facilitate a path toward a resilient future. In this paper, experiences from Ireland in terms of how we might envision sustainable rural landscapes going forward are explored with the aim of outlining ways in which policies dealing with landscape management practices, particularly as they relate to farming, can be more carefully constructed and result in successful rollout and acceptance by those charged with operationalizing them on the ground. What is also evident is that the underpinning concept that is referred to as a sustainable landscape needs to be appreciated in terms of its dynamism and continual alteration, as well as how its meaning, significance and management changes over time [1]. This paper then engages with farmers on the ground to identify how it is possible to develop 'intelligent decisions' [2] about the future of rural landscapes and, while not claiming to be the only pathway, presents an approach that demands serious consideration in terms of sustainable landscapes and what that might mean going forward.

The interesting thing about placing this research in the Irish landscape is its mosaic of topographies—rich agricultural lands, mountainous disadvantaged areas and internationally recognised Special Areas of Conservation—a past of over-exploitation, degradation and pollution and, in more recent times, a re-emergence of a new appreciation for protection and management of rural landscapes and their innate fragility. This paper, contextualized by Selman's dimensions of sustainable landscapes and particularly those focused on environment, economic, social and governance aspects [2] explores three aspects of recent Irish landscape management practices, namely that of Designation, whereby lands are declared as Special Areas of Conservation (SACs) (see Table 1 for a list of acronyms used in the text) and all that that entails; Appreciation—where some recent initiatives have sought to engage more closely with local farmers and allow their knowledge and input to influence how rural landscapes are managed; and lastly, Incentivisation, where results-based payments play a central part in directing farming toward more environmentally friendly practices and management systems. What emerges is that, while impacts vary, it is in combination, under locally led decision-making models, that the best solutions are found.

Table 1. List of acronyms.

AES	Agri-Environmental Schemes
EIP	European Innovation Partnership
EIP-AGRI	European Innovation Partnership for Agricultural productivity and Sustainability
GLAS	Green Low-Carbon Agri-Environment Scheme
IFDL	Irish Farmers with Designated Land
LLAES	Locally Led Agricultural Environmental Scheme
NPWS	National Park and Wildlife Services
SAC	Special Area of Conservation

2. Contextualising Rural Landscape Management

The rural occupies a central place in the European narrative. At its most basic, the agenda of providing safe and quality food and the expectation of a reasonable standard of living is constant, couched as it is in broader acceptance of sustainability principles and the need to be mindful of the needs of future generations. Indeed, McDonagh [3] argues that 'Europe and beyond are struggling with the twin challenges of producing safer quality foods, while preserving, if not enhancing, the natural environment in which it is produced' (p. 6). In terms of the crossroads that we find ourselves at currently, the decision of which direction to take is fraught with complexity. Somewhat similar to the ease with which support can be proclaimed for the principles of sustainability, it is less clearcut when it comes to enacting measures on the ground. This is particularly so when meeting the needs of the present can often outweigh the necessity to make unpalatable decisions that are for future well-being. Nowhere is this epitomised more than in our rural landscapes. Rural areas provide the ingredients for human survival, yet they are also those most threatened by human activity. Indeed, the management of rural landscapes has never been straightforward and can best be thought of as vacillating from extremes of great concern to that of reckless overuse [4].

Sustainability, as a consequence, has become the central tenet of contemporary policies relating to the use and management of rural resources. Pressure in recent years has ratcheted up considerably in terms of demands for a more sustainable management trajectory for rural areas and its role in issues such as climate change, in particular. Agriculture and agricultural practices have, perhaps, the most profound impact on the myriad of habitats and species that are found across our rural areas [5–8]. In fact, it could be argued that there is a strong interdependency between the protection and conservation of habitats and species and the type and intensity of the farming systems engaged in it. What is equally valid is that, in recent decades, the incongruity that is farm intensification and land abandonment has dramatically altered our biodiversity and broader ecosystems [8]. Indeed, it took the economically unsustainable and politically untenable situation of continuing to subsidise farmers' augmentation of existing overproduction to bring about adjustments to EU agricultural policy [6]. This brought forth incentives to promote not only pluriactivity, particularly in the form of on-farm diversification, but also efforts to encourage more environmentally friendly farming practices. This, for the first time, led to measures for financial benefits being provided to farmers who were willing to adapt their farming to nature conservation requirements. This change in mindset was a major step forward, although it can be somewhat tempered in that farmers that availed of these measures were often those already conducting environmentally friendly farming practices in the first place.

In all, while initially in small steps, the greening of the agricultural agenda began with the objectives of making farming both multi-functional and profitable. A key component, and what has often been described as one of the most impressive achievements in the environmental field, was the emergence of agri-environmental policies [7–9]. What was apparent here was how the decline in extensive or traditional farming practices had placed a downward pressure on habitats and rural biodiversity. The drive to produce more had seen agricultural outputs reach unprecedented levels with consequent surpluses being put into intervention or destroyed [6]. While this continued unchecked for a number of years, in recent times, and particularly in the last decade, we have seen efforts to realign where agriculture is going, prompted by a growing concern for habitats and species decline and climate change. Indeed, the European agenda has, if not quite pivoted, reaffirmed its desire to one of 'attempt(ing) to manoeuvre agriculture to a more acceptable position with farmers being cast as custodians of the countryside' [10] (pp. 713–714). There have been a number of ways in which this realignment has played out on the ground. Undoubtedly, agri-environmental policy has been fundamental in creating numerous positive environmental responses to agriculture, including growing knowledge of nutrient and habitat management and educating toward greater environmental awareness. This process was not without its critics, however. The European Court of Auditors, for one, called for more targeting of agri-environment payments, an observation that was heeded in the development of Locally Led Agricultural Environmental Schemes (LLAESs), which bring a much-needed impetus to landscape management practices and a contribution that is examined later in this paper.

3. Ireland's Approach to Landscape Management

Ireland is a prime example of the challenges facing farmers and their sustainable management of the resources on which their livelihoods depend. Pressures associated with agriculture have had major impacts on land-based habitats and species, with over 70% of the number of habitats of EU interest reported to be negatively impacted by agriculture [5]. Ecologically unsuitable grazing regimes and abandonment are the main Irish pressures reported, with the Irish National Biodiversity Plan 2017–2021 [5] declaring that the breeding populations of bird species that are associated with farmland, such as the Curlew, Lapwing and Yellowhammer, have declined substantially over recent decades, some to the brink of extinction [10]. From such a starting point, the challenge in terms of conservation being integrated into sustainable agricultural practices is significant.

Intensification, a key aspect of the modernisation of agriculture, had the unfortunate side effect of increasing pressure on the environment. Consequently, the reforms of the EU's Common Agricultural Policy (CAP) since 1992 have aimed to progressively reduce this pressure with several instruments and tools developed to mitigate the environmental impact of agriculture. Agricultural Environment Schemes (AESs) have been one of the foremost of these tools. Developing alongside the newfound interest in AES, a growing number of protected areas also emerged globally. As well as their development, what is also interesting is the way in which many of these areas are governed and managed, particularly as they are integral in bringing together environmental concerns and the practice of farming.

3.1. Special Areas of Conservation (SAC)

There are over 13,500 sq. km of SAC designations in Ireland covering all types of landscapes. The designation of SACs stems from the EU Habitats Directive, and they are part of the NATURA 2000 network of European protected sites. There are over 400 SAC designated sites in Ireland covering 13% of the land area [10]. The challenge for landowners, even if they want to protect the habitats on their land, is that they are not automatically entitled to be compensated for SAC being part of their holdings. They are, however, restricted in the type of activities they can engage in and are fined if they breach SAC restrictions. While a vast majority of SACs in Ireland are state owned (by the National Parks and Wildlife Service (NPWS), Coillte (Forestry Agency) and the ESB (Electricity Supply Board), for example), there are many farmers who are affected by this designation process. In one effort to address this, a National Farm Scheme was introduced in 2004 that provided compensation to SAC designated farms, but its decline in 2012 led to farmers being left with the 'burden' of SAC designation and being 'unable to farm as they wished and ... unable to maximise their lands to their full potential' [11,12]. What is significant in this process, and a situation that undermines the role of farmers somewhat in terms of their role in managing the landscape, is how SAC designation is very much a top-down political and scientific decision-making process. Landowners are notified of the proposed designation, are sent an information pack explaining the scientific reasons for the designation and are given details of how they can go about appealing the process. What is readily apparent is the absence of any attempt at consultation between landowner and policymaker. This top-down approach excludes input from the farmer, leaving a situation of disconnect and disempowerment rather than trust and mutual respect. Discussions with farm owners reinforce the view of limited participation, which, if it did occur, came 'after scientific evaluations and decisions have been made (and carries) a strong top-down "conservation" imprint with less regard for its social acceptance and feasibility at a local level, although land designated is to be managed by farmers' [13] (p. 29). Indeed, the designations of SAC sites has often met with opposition from groups such as the Irish Farmers Association (the IFA is the largest farmer organisation in Ireland), with even a dedicated group called the Irish Farmers with Designated Land (IFDL) being set up. The aim of this group was one of uniting farmers and landowners in regaining the value of designated land and ensuring farmers could generate a reasonable income from designated lands [14].

3.2. Locally Led Agricultural Environment Scheme (LLAES)

One programme which has the potential to address the shortcomings of Designations and the broader remits of national Agricultural Environmental Schemes is that of the Locally Led Agricultural Environmental Scheme (LLAES). Prompted by EU policy, such schemes are intended to encourage locally driven solutions to address local issues. Theoretically, these schemes present a real opportunity to reach the spaces where knowledge is shaped and transferred. Ireland's Locally Led Agricultural Environment Schemes (LLAESs) are specifically targeted to meet the requirements of EU Birds, Habitats and Water Framework Directives and aim to address particular environmental and biodiversity challenges not addressed at national level by the Green Low-Carbon Agri-environment Scheme (GLAS) (this scheme was introduced under the Irish Rural Development programme 2014–2020 and provides payments to farmers to help tackle climate change, preserve biodiversity, protect habitats and promote environmentally friendly farming) [15]. The centrally identified priorities include the continuation of the BurrenLIFE programme, priority pearl mussel catchments and hen harrier areas. LLAESs encourage locally driven solutions and require submission of proposals by local groups accompanied by detailed estimates of costs. A great example is evident in the Burren Programme. This innovative programme takes a farmer led approach, where the farmer nominates and co-funds conservation actions on their farm, giving the farmer a type of freedom to farm. What makes the Burren different is that it combines these actions with a results-based payment. To ensure that the desired results are achieved, payments are made to farmers based on the environmental condition of their farm. The better a farmer's field 'scores' in terms of environmental outcome, the more payment they receive. The way in which these scores are reached and implemented is that "eligible field are assessed annually by the farm advisor using a user-friendly "habitat health" checklist. Farmers are made aware of their scores (and) all scores are reviewed for accuracy and consistency by the Burren team and many also are checked by Dept. of Agriculture inspectors. The field score, which ranges from 0 to 10, is calculated using nine distinct, weighted criteria which, taken together, give a very accurate picture of the "health" of the grazed habitats in that management unit. These criteria are: Grazing level; Amount of litter (dead vegetation); Extent of feed site damage; Extent of damage at natural water sources; Level of bare soil and erosion; Level of encroaching scrub; Amount of bracken and purple moor grass; Extent of weeds and agriculturally-favoured species; and Ecological integrity. Once the field score is calculated, it is multiplied by the available payment rate per hectare and by the size (ha) of the field, to calculate the "output payment" for that field. Under the Burren Programme, all fields with a score of 6 or more receive payment but higher scores receive higher payments—increased payment rates are available for fields scoring 9s and 10s. Fields with a score of 5, only receive payment in the first two years. Payments per ha can range from €8/ha to €180/ha depending on field score and farm size (payments are "banded" to reward smaller holdings). This gives farmers the incentive to manage their fields in ways that will improve their scores and their payment as well" [16].

A key component is the way in which there is a partnership approach involving all the key stakeholders, farmers, state agencies and government departments in tailoring solutions in practical and applied ways.

Indeed, it has been suggested that LLAESs could be rolled out across EU Member States and variously adapted and applied to areas of High Nature Value (HNV) throughout [17,18]. This type of approach is now beginning to emerge as a key instrument in the new CAP 2023–2027, with Ireland, for example, instigating a new Results-Based Environment Agri Pilot (REAP), which seems to be generating a lot of interest from farmers, with 10,000 initial applications received thus far [19].

3.3. *EIP and EIPAgri*

The emergence of European Innovation Partnerships (EIPs) has stemmed from the broader EU strategy of 'Europe 2020', with its desire to ensure that EU citizens have increased employment opportunities, a better quality of life and a competitive economy in which to live [20]. The basic tenet is that of bringing together public and private sectors across all scales to co-operate through research, innovation and practice in building a better and more sustainable and inclusive economy. EIP-Agri, launched in 2012, focused on bringing together stakeholders engaged in agriculture and forestry sectors, with innovation and fostering co-operation between researchers and practitioners being fundamental. In particular, EIP-Agri has at its core Operational Groups that explore new insights while also drawing on existing tacit knowledge in a bid to find solutions to specific issues (agricultural and forestry sectors) and to develop new opportunities (ibid). Essentially, what makes this initiative a step in a different direction is that it provides funding (€59m) to bring together as many stakeholders as possible, with the aim of dealing with a local issue locally [20]. When EIP-Agri was launched in Ireland in 2016, there was a large volume of interest and a large array of projects and proposals submitted. There are currently 23 projects operating on the ground, and their impact has been hailed a significant contribution in how we manage our rural landscapes. Indeed, EIP-AGRI projects have been described as being central to addressing 'challenges such as biodiversity, profitability and sustainability (while) harness(ing) the creativity and resourcefulness which is the hallmark of Ireland's rural sector' [21] (p. 3).

3.4. Results-Based Payments

There is a shift to move beyond compensating farmers for halting negative practices and instead incentivise positive management by paying for the delivery of clearly defined, measurable environmental outputs/results. Wynn-Jones [22] suggests that results-based agri-environment schemes should be seen as a 'new form of production' (p. 77) rather than an offshoot of agriculture. In essence, the results-based approach looks to instil behavioural change in what a farmer does through adapting to the specific demands of different areas and through famers being centrally involved in the design and development of how environmental objectives should be attained. This approach instils a greater sense of ownership among farmers and provides a strong platform on which environmentally positive farming practices can be built. The payment structure of agri-environment programmes can be divided into two main categories: outcome/results-based payments and prescription/action-based payments. Research and results from a number of pilot projects have shown that results-orientated agri-environment programmes offer a more effective means of delivering better environmental outcomes if they are well designed and are accompanied by robust environmental indicators to measure outcomes [23,24].

The positivity that is growing around a results-based payment system centres on how it allows farmers greater freedom to decide how to manage their land and use their skill and experience to improve environmental and agricultural performance. In addressing broader policy concerns, it also suggests a better-value-for-money model, in that, if deliverables are not met, then payment is not made. While one of the weaknesses of the action-based system is the lack of monitoring or ability to measure positive changes, results-based payments employ a field scoring system which generates data that help determine where positive environmental impacts are occurring [16]. Undoubtedly this approach requires a change in mindset of farmers toward how they farm. However, what is also apparent is that this new pathway very much calls into question the 'one-size-fits-all' approach favoured in EU policy to one that takes cognisance of location-specific needs and challenges. This approach not only demands flexibility and adaptability but draws on and incorporates farmer/local knowledge into its design and development. Consequently, what has emerged in the recent literature is that results-based approaches can not only add value to action-based ones, but 'can be adapted to complement action-based approaches and be geographically targeted to situations where they are best suited' [25] (p. 296).

4. Methodology

As well as drawing on broader research concerning aspects of land-use management, empirical evidence for this paper was also drawn from a series of interviews conducted in the Western Region of Ireland. The information gathered greatly enhanced an understanding of the broader discourse in the area of land management, use and protection. In all, 30 semi-structured interviews took place over a period of months during 2017/2018, and these were backed up by other documentation, including website materials, brochures and newspaper articles. The interviews were recorded and transcribed, and when used in the text, they are indicated by letter and number. Nvivo was used in relation to the data analysis, alongside a content analysis approach. The Western Region of Ireland affords insight into an unusual amalgam of land use, economic activity, conservation and sustainability (economic and social) challenges. The interviews with the landowners allowed us insight into and access to first-hand knowledge of, and dealings with, fragile landscapes and the farming practices therein. Two areas in the western part of Ireland, namely the Burren (comments from those interviewed are identified by the letter B and a number) and the region around the Ox Mountains (comments from those interviewed are identified by the letters Ox and a number), were the main study areas. The Burren is a particularly fragile landscape that is noted for its archaeology, flora and fauna interspersed by a considerable amount of farmed land, while the Ox Mountains has the typical challenges that come with a mixed topography, fragmented holdings and mixed-quality lands in terms of farming practices. The farmers talked to us, providing an array of viewpoints, with some being more advanced in their own engagement with sustainable farming practices, that is, the various AES schemes available, than others.

5. Results and Analysis

In recent decades, various approaches have been made to halt the decline of biodiversity in agricultural areas; designating lands as protected areas and providing financial incentives to farmers to join agri-environmental schemes have been typical instruments.

5.1. Designation

In terms of land designation, the process has often met with conflict from landowners and disillusionment at the perceived top-down lack of consultation that this process has often entailed. The reaction to land designation in Ireland (and elsewhere) is often projected as a feeling of exclusion from the procedure, with farmers citing the absence of local input, knowledge, traditions and values in the drawing up of designations [26,27]. Consequently, many designations often result in, if not failure, only lukewarm acceptance. In the context of this study, the empirical results reinforced many of these aspects, with top-down decision-making, the complexity of the process and the type of prescriptive type of payment process all coming in for criticism.

SAC designation has tended to reflect top-down attempts at participatory or multi-stakeholder consultation and never fully embraced local knowledge and practices as valuable expertise in the sustainable management of these areas [28]. During the course of the fieldwork for this paper, there was undoubted support and understanding of the need to protect fragile landscapes and acknowledgement of historical, cultural and environmental significance in respective regions. However, the cursory inclusion of farmers as key stakeholders, as well as their peripheral placement in the design and development of SACs, engendered what could best be described as a sense of disillusionment. Indeed, in terms of land designations, there was a very emotional response in the interviews to this perceived lack of involvement, as well as a sense of frustration. Some farmers pointed toward the absence of local knowledge and the role it could play with one declaring that *'they're a bit useless for some farmers'* (Ox18), while another suggested that *'they don't make the most of what could be done on different areas of land. I don't have great things to say about them anyway'* (Ox8). There was also a sense of anxiety evident with regard to how restrictions were placed on private lands which conflicted with a farmers' desire to meet their own land-use objectives: *'There's another one near us, another SAC ... I wouldn't like to see us being in one really because it would upset the land'* (Ox1). The additional costs associated with designation were also highlighted: *'It's a bit of a problem I'd say, especially young people going looking for planning permission now because it becomes an awful lot more expensive you have to get an archaeological report on the site, so it's a big problem'* (B1). Others referred to their own farm enterprise and the knock-on effect such designation on their lands would have in that there would be costs associated with not being able to develop their farms and *'make a decent living'* (B2), because, as one farmer also pointed out, *'you see ... I can't really see how I can improve ... if you compare ... to somebody that isn't on the (SAC) site, they can do what they want so of course there's an economic impact'* (B1).

It was also felt by many of those interviewed in both study regions that many existing farming systems and practices were already compatible with environmental goals and that some of the new schemes were going against what they felt worked best: *'all of these rules now don't apply to the reality, they don't account for best practices'* (Ox3). One farmer pointed to how *'some of the schemes don't really understand what we're about'*, while another referred to the knowledge that was handed on to him and which was now being ignored: *'one of my uncles taught me all I know about this mountain, what time of year to put sheep out. Now for GLAS we're forced to keep them out for seven months of the year ... they don't want to know the fact that the grass only turns sweet at the end of May beginning June on a good year'* (Ox8). Many of the farmers interviewed pointed to difficulties with the top-down nature of the process and what they perceived as the *'endless form filling required'* (B7). In some of the conversations during the field research, farmers suggested that *'these schemes need to be simple and flexible if they want farmers to buy into them'* (Ox4) and they *'need to suit the area'* (B6).

While the consensus among the farmers saw merit in what designations were trying to achieve, they also felt that, as the ones charged with delivering on these objectives, they had very little input into the design or decision-making around the process: *'we're the ones that should be front and centre ... we're doing all the work'* (Ox2). This was evident with farmers critical of the lack of dialogue between top-down and bottom-up in terms of existing policies and practices, with a farmer declaring that *'it's one way ... there's no question of tell us back what does and doesn't work for you. It is very much a one way street'* (Ox3). Another farmer reinforced this sentiment in his comment that there was a *'big disconnect between what the policy makers think they know and what was is really happening on the ground'* (B8). While the issues of payments and prescribed rules (and penalties) were a common thread in many of the discussions, it was also clear that the farmers wanted to be heard and appreciated for the knowledge and experience they themselves had accrued, as signified by the comments of one farmer who declared the following: *"if it was farmed with some of the practices that I know it would respond better and I'm under no doubt about that"* (Ox7).

The payment structure of the AES has often proven to be problematic in that it is based on prescription/action-based payments for the adoption of particular land uses or land management practices [29]. These scientifically defined criteria with prescribed sets of rules [17] do not, for the most part, account for local conditions or farmer knowledge. Consequently, frustration, if not anger, was evident during the field study in relation to the perceived contradictory nature of some policies and the subsequent knowledge that farmers receive. The often-punitive nature of the policies, that is, punishment for wrongdoings rather than incentivise for good, often left farmers frustrated. One farmer recounted how *'one fella ... took over a farm here and he was doing improvements on it and they stopped him, and ... they stopped his payments ... he was mak(ing) better walls ... but they wouldn't let him do that'* (B1). Another farmer referred to how there was a lack of communication between policymakers and what a farmer could engage with on the ground: *'you see ... penalties were imposed on the basis of insufficient farming activity, but the Department (of Agriculture) never set out a criteria for what they defined as insufficient farming activity, neither did they take into account the environmental condition of these lands or even carry out an appropriate assessment where lands had a designation on them'* (Ox7).

The example of the Burren programme is perhaps the catalyst for change that is necessary to ensure greater buy-in and support from farmers; many of the procedural obstacles were removed through innovative processes such as unique field scoring systems and simplified farm plan and paperwork, which all help 'minimise the bureaucratic burden' [16]. This is essentially, the provision of a scheme that focuses on conservation results rather than strict management methods or prescriptions, as well as being one that is tailored for areas with differing farming systems, habitats and species.

5.2. Incentivisation and Participation

Studies show that polices tailored to the situation on the ground, with adjusting payments, as well as financial incentives, have positive influences on participation [30]. In the context of agri-environmental schemes, the approach has been very often couched in the belief that farmers' actions were primarily driven by an economic rationale, and, therefore, providing economic gains for farmers would lead to a change is practice [31]. The reality, however, proves somewhat different, with many farmers in the study areas being multi-generational and deeply imbedded in their rural communities and societies. While financial incentives may 'buy' some commitment, as one farmer commented, *'you don't farm in area like this for the money'* (B2).

The concept of Locally Led Agricultural Environmental Schemes (LLAESs) has, therefore, significant potential to bring innovative solutions to bear and to ensure sustainable land management. Likewise, EIPs promote local solutions to specific issues and involve the establishment of Operational Groups (OGs) to develop ideas or take existing ideas/research and put them into practice by being hands on in terms of working toward the resolution of a practical problem. The results from the Burren Programme reinforce these observations with clear evidence of direct and indirect social, economic and environmental impacts being accrued. Farmers talked of the freedom they have been given whereby they use their own skills and experience to deliver on environmental needs, in addition to their farms benefitting from better management practices. This was also very much reflected in the comments of a farmer from the Ox Mountains region when he declared that '*farms should continue to be economically viable for the people farming on Ox mountain like it has been for generations*' (Ox7).

One of the basic principles of the Burren Programme was that the learning process was based on the participation of farmers in the process. They were the ones experimenting, evaluating and selecting practices and solutions that adapt best to their own farm's conditions. In this process, local tacit knowledge is not arbitrarily extracted; it is shared knowledge. Scientists, technicians and farmers work together and are coordinated through horizontal linkages. The best outcome and best possibility of getting buy-in from farmers seems to rest on their meaningful inclusion, and, as reflected in the comments of one farmer, they also '*want the area to be protected . . . I don't want an area to die . . . I want it to be a lived-in landscape*' (B1). This sentiment was also reflected in the comments of a farmer from the Ox Mountains region who suggested that '*you need to have all voices heard*' (Ox7). In place of generic criteria imposed through government policy and farmers in engaging in, what one described as '*a guessing game*' (Ox3), having farmers involved in the decision-making process from the initial design stage through to how each plan is derived allows the farmer to develop a sense of ownership and certainly has proven to deliver results [32].

The outcome-based approach sees the programme manager paying for results and, hence, not looking for breaches, ensuring a better working relationship and fewer non-compliance issues [17]. A devolving of power to farmers in terms of self-assessment on how they are preforming also has significant impacts in terms of the level of trust between top-down and bottom-up. In addition, a 'personalised' designing of individual farm plans, with advisor and farmer working hand in hand, sees the farmer's knowledge being greatly valued and given status through its incorporation into the design of their farm plan. All of these measures are extremely positive steps going forward in terms of addressing the issues of disconnect felt between farmers and policymakers and the sense of exclusion felt by farmers in terms of being able to contribute to policies that they are expected to deliver on. This type of integrated strategy was described by one farmer as being 'essential' and a necessary requirement to '*drive that forward . . . so you can create and implement structures that will foster local communities, and local communities here (are) centred around farming and that shouldn't have to change*'(Ox7).

6. Discussion and Concluding Remarks

During the course of this discussion, a number of aspects have come to the fore, and while the evidence presented here reflects the possibilities within an Irish landscape, the message of positioning a farmer's input and knowledge in a prominent position in the designing of agri-environmental programmes is applicable across Europe and beyond. While by no means extensive, the following have thus emerged as key ingredients in the pursuit of sustainability pathways for agriculture in the coming decades:

6.1. The Importance of Multi-Stakeholder Involvement and a Prominent Role for Farmers in the Decision-Making Process

There is a realisation that, while direct intervention can have a specific impact in the drive toward biodiversity conservation, the inclusion of farmers is an essential component. While the propensity has been for a top-down scientific driven model, multi-stakeholder involvement is paramount. The willing participation of farmers, who, for the most part, are the main landowners, and a sense of ownership of policy measures that impact their practices, is vital to the effective implementation of any landscape management approach.

This was very evident in the practices found in the Burren, where there was a strong rapport between stakeholders with the input of the farmer carrying weight and value. The conversations and data from the Ox Mountains area, however, suggested that there was still some way to go to achieve such inclusion and partnership. Indeed, it could be argued that the Burren is more the exception with the Ox Mountains region more reflective of the broader challenges still to be addressed. In particular, many of the reflections and discussions with the farmers suggested indications of disconnect between the various stakeholders and the vision that was trying to be moulded. In fact, the integration of local experiences, scientific knowledge and farmers' ideas into policymakers' demands was in its infancy in terms of the requisite trust and partnership required. In many comments, there seemed to be a sense of powerlessness among local landowners in terms of the decisions being made about how the landscape should be managed and a feeling that there was negligible recognition of their role, with continued debates on who, when, where and what knowledge are included [33]. Although national AES remains a science-first or ecology-first process [34], the LLAES demonstrates a potential to provide a platform on which scientists, policymakers and farmers can work toward a common goal in terms of better environmental practices and land stewardship.

6.2. *The Combination of Action-Based, Results-Based and Locally Led Programmes*

The demands for environmental protection and management have never been greater. The necessity to ensure this needs the appropriate architecture that will enable such a pathway to evolve. The combination of action-based and results-based approaches, allied to locally adapted practices, can drive this change. This combination will invariably enhance environmental practices. Perhaps the most striking aspect of this new pathway very much calls into question the 'one-size-fits-all' policy direction currently dominating, to one that takes cognisance of location-specific needs and challenges. While the specific attributes of any given place demand a certain type of approach, there are many commonalities that can be operationalised across all EU member states. The combination of action-based, results-based and locally led programmes provides flexibility and adaptability, while also drawing on and incorporating local farmer knowledge in design, development and implementation. The Burren Programme reflected this particularly well in that it adopted a hybrid outcome-based payment system, with two main measures that absorb roughly equal funding, one for actions (capital works) and the other for outputs/results [32]. The positive outcomes thus far point to a process that enables an alternative means of achieving environmental objectives.

6.3. *Integrating Local and Scientific Knowledge in Pursuit of the Best Environmental Outcomes*

If we accept that 'knowledge-sharing and community-learning processes ... contribute to sustainability' [35] (p. 257), then the absence of such practices is surely undermining efforts at developing a sustainable future. McDonagh et al. [13] suggested that there is 'often (a) contradictory nature of top-down policies that frustrate those on the ground and in many instances create unnecessary tension and conflict' (p. 122). The challenge for policymakers is one of how to engage local people and how to extract their 'tacit and embedded knowledge' [36] and not in a 'box-ticking' way or one that does not 'recognise the value this can add to decision-making related to landscape and natural resource use' [13] (p. 126). Indeed, it is hard to envisage the landscapes of the two study regions explored in this paper being sufficiently maintained without 'the land management and livestock husbandry skills of farmers and the cultures of their communities' [37] (p. 90). Consequently, the important role played by the farmers' experience and the knowledge they possess cannot, and should not, be dismissed or underestimated. Indeed, a key aspect of the activities in the Burren region is very much about demonstrating to, informing and listening to farmers instead of imposing restrictions. The Burren Programme, for example, consciously incorporated these insights into its programme, with locals and part-time farmers given extensive training and then being hired as advisors to work for the Burren Programme.

This has the important outcome of developing a sense of ownership within the community and, more important, increasing the connection between landowner and policy objectives.

In a final comment, there is no doubt that the decline in global biodiversity is at a critical level, and the reduction in the biodiversity on our farms is a major contributor to this. What has been presented here is the significant role that carefully crafted and inclusive agri-environment schemes can deliver. An opportunity which can be a key instrument is addressing biodiversity decline and one that can fashion a pathway toward greater resilience and sustainability in our rural landscapes.

Funding: This research received no external funding.

Institutional Review Board Statement: Not applicable.

Informed Consent Statement: Not applicable.

Data Availability Statement: Not applicable.

Conflicts of Interest: The author declares no conflict of interest.

References

1. Antrop, M. Sustainable landscapes: Contradiction, fiction or utopia? *Landsc. Urban Plan.* **2006**, *75*, 187–197. [CrossRef]
2. Selman, P. What do we mean by sustainable landscape? *Sustain. Sci. Pract. Policy* **2008**, *4*, 23–28.
3. McDonagh, J. Discourses of food and Sustainable Rural Futures. *Prog. Hum. Geogr.* **2014**, *38*, 838–844. [CrossRef]
4. Murray Gray, J. Planning and Landform: Geomorphological Authenticity or Incongruity in the Countryside? *Area* **1997**, *29*, 312–324. [CrossRef]
5. Irish National Biodiversity Plan. Available online: www.npws.ie/legislation/national-biodiversity-plan (accessed on 23 February 2022).
6. Dax, T.; Kahila, P. Policy Perspective—The evolution of EU Rural Policy. In *The New Rural Europe: Towards Rural Cohesion Policy*; Copus, A., Hornstrom, L., Eds.; Nordregio: Stockholm, Sweden, 2011.
7. Potter, C. *Agri-Environmentalism and Rural Change*; Elsevier: Amsterdam, The Netherlands, 2009.
8. Van Herzele, A.; Gobin, A.; Van Gossum, P.; Acosta, L.; Waas, T.; Dendonker, N.; de Frahan, B.H. Effort for money? Farmers' rationale for participation in agri-environment measures with different implementation complexity. *J. Environ. Manag.* **2013**, *131*, 110–120. [CrossRef] [PubMed]
9. Robinson, G. *Geographies of Agriculture*; Routledge: London, UK, 2004.
10. McDonagh, J. Changing expectations and contradictions in the rural. *Prog. Hum. Geogr.* **2013**, *37*, 712–720. [CrossRef]
11. NPWS (National Parks and Wildlife Services). Available online: https://www.npws.ie/protected-sites/sac (accessed on 14 January 2022).
12. Kiernan, A. Calls for Reinstatement of the National Farm Scheme for SAC Farmers. 2019. Available online: https://www.agriland.ie/farming-news/calls-for-reinstatement-of-the-national-farm-plan-scheme-for-sac-farmers/ (accessed on 14 January 2022).
13. McDonagh, J.; Olafsdottir, R.; Weir, L.; Mahon, M.; Farrell, M.; Welling, J.; Conway, T. There's no transfer of knowledge, it's all one way—the importance of integrating local knowledge and fostering knowledge sharing practices in natural resource utilisation. In *Sharing Knowledge for Land Use Management*; McDonagh, J., Tuulentie, S., Eds.; Edward Elgar Publishing: Cheltenham, UK, 2020; pp. 116–129.
14. Irish Farmers with Designated Land. Available online: https://data.oireachtas.ie/ie/oireachtas/committee/dail/32/joint_committee_on_communications_climate_change_and_natural_resources/submissions/2018/2018-01-30_opening-statement-jason-fitzgerald-irish-farmers-with-designated-land-ifdl_en.pdf (accessed on 23 February 2022).
15. DAFM (Department of Agriculture, Food & Marine). Available online: https://www.gov.ie/en/service/9133a5 green low-carbon-agri-environment-scheme-glas/# (accessed on 28 April 2022).
16. Burren Programme. Available online: http://burrenprogramme.com/impact/outputs/ (accessed on 10 December 2021).
17. McGurn, P.; Moran, J. *A National Outcome Based Agri-Environment Programme under Ireland's Rural Development Programme 2014–2020*; Report produced for the Heritage Council; Atlantic Technological University Sligo: Sligo, Ireland, 2013.
18. Keenleyside, C.; Radley, G.; Tucker, G.; Underwood, E.; Hart, K.; Allen, B.; Menadue, H. *Results-Based Payments for Biodiversity Guidance Handbook: Designing and Implementing Results-Based Agri-Environment Schemes 2014–2020*; Institute for European Environmental Policy: London, UK, 2014.
19. Carty, D. Individual entry options for new agri-environment schemes. *Ir. Farmers J.* **2022**. Available online: https://www.farmersjournal.ie/individual-entry-options-for-new-agri-environment-schemes-681660 (accessed on 27 February 2022).
20. Europa.eu. Farming: Good Management Practices for Natura. 2000. Available online: https://ec.europa.eu/environment/nature/natura2000/management/gp/farming/02case_termoncarragh.html (accessed on 15 November 2021).
21. Conway, S.F.; Farrell, M. *EIP-AGRI: Ireland's Operational Groups 2019*; Department of Agriculture, Food and the Marine (DAFM): Dublin, Ireland, 2019.

22. Wynn-Jones, S. Connecting payments for ecosystem services and agri-environment regulation: An analysis of the Welsh Glastir Scheme. *J. Rural Stud.* **2013**, *31*, 77–86. [CrossRef]
23. Matzdorf, B.; Lorenz, J. How cost-effective are result-oriented agri-environmental measures?: An empirical analysis in Germany. *Land Use Policy* **2010**, *27*, 535–544. [CrossRef]
24. Osbeck, M.; Schwarz, G.; Morkvenas, Z. *Dialogue on Ecosystem Services, Payments and Outcome-Based Approach*; Baltic Compass: Uppsala, Sweden, 2013.
25. Finn, J.A. Synthesis and reflection on selected results-based approaches in Ireland. In *Farming for Nature—The Role of Results-Based Payments*; O'Rourke, E., Finn, J.A., Eds.; Teagasc and National Parks & Wildlife Service (NPWS): Dublin, Ireland, 2020.
26. Bryan, S. Contested boundaries, contested places: The Natura 2000 network in Ireland. *J. Rural Stud.* **2012**, *28*, 80–94. [CrossRef]
27. Holmgren, L.; Sandström, C.; Zachrisson, A. Protected area governance in Sweden: New modes of governance or business as usual? *Local Environ.* **2017**, *22*, 22–37. [CrossRef]
28. O'Rourke, E. Socio-natural interaction and landscape dynamics in the Burren, Ireland. *Landsc. Urban Plan.* **2005**, *70*, 69–83. [CrossRef]
29. Matzdorf, B.; Kaiser, T.; Rohner, M.-S. Developing biodiversity indicator to design efficient agri-environmental schemes for extensively used grasslands. *Ecol. Indic.* **2008**, *8*, 256–269. [CrossRef]
30. Raggi, M.; Viaggi, D.; Bartolini, F.; Furlan, A. The role of policy priorities and targeting in the spatial location of participation in agri-environmental schemes in Emilia-Romagna (Italy). *Land Use Policy* **2015**, *47*, 78–89. [CrossRef]
31. Burton, R.J.F.; Paragahawewa, U.H. Creating culturally sustainably agri-environmental schemes. *J. Rural. Stud.* **2011**, *27*, 95–104. [CrossRef]
32. Dunford, B. The Burren Life Programme: An Overview. Research Series Paper No. 9 March 2016. NESC. Available online: http://files.nesc.ie/nesc_research_series/Research_Series_Paper_9_BDunford_Burren.pdf (accessed on 15 November 2021).
33. Pinton, F. Conservation of biodiversity as a European directive: The challenge for France. *Sociol. Rural.* **2001**, *41*, 329–342. [CrossRef]
34. Stoll-Kleeman, S.; O'Riordan, T. From participation to partnership in biodiversity protection: Experience from Germany and South Africa. *Soc. Nat. Resour.* **2002**, *15*, 161–177. [CrossRef]
35. Sisto, R.; Lopolito, A.; Van Vilet, M. Stakeholder participation in planning rural development strategies: Using backcasting to support Local Action Groups in complying with CLLD. *Land Use Policy* **2018**, *70*, 442–450. [CrossRef]
36. Kelemen, E.; Megyesi, B.; Kalamasz, I.N. Knowledge dynamics and sustainability in rural livelihood strategies: Two case studies from Hungary. *Sociol. Rural* **2008**, *48*, 257–273. [CrossRef]
37. Condliffe, I. Policy change in the uplands. In *Rivers of Environmental Change in Uplands*; Boon, A., Allott, T., Hubacek, K., Steward, J., Eds.; Routledge: London, UK, 2009.

MDPI

Article

Moravian–Slovak Borderland: Possibilities for Rural Development

Antonín Vaishar [1], Milada Šťastná [1,*] and Hilda Kramáreková [2]

[1] Department of Applied and Landscape Ecology, Mendel University in Brno, Zemědělská 1, 61300 Brno, Czech Republic; antonin.vaishar@mendelu.cz

[2] Department of Geography and Regional Development, Faculty of Natural Sciences, Constantine the Philosopher University in Nitra, Trieda Andreja Hlinku 1, 949 74 Nitra, Slovakia; hkramarekova@ukf.sk

* Correspondence: milada.stastna@mendelu.cz; Tel.: +420-54513-2459

Abstract: This article analyzes the question of how the change of geopolitical position in the rural region of Eastern Moravia, which was shifted from the center of the state on its border, is reflected. The paper shows how the originally marginal region transformed from an area with shepherd agriculture to an industrial area with a skilled workforce during the existence of Czechoslovakia and questions how to cope with the consequences of the reverse change into a marginal geopolitical position on the eastern border of Czechia. The paper considers the balance of migration, supplemented by the construction of new dwellings, to be a relatively complex indicator. It states that the region of Eastern Moravia is problematic in terms of further development, except for the northern part, which is affected by the suburbanization of Ostrava. As a result, it proposes to supplement the current orientation toward the manufacturing industry by creating conditions for the development of cultural tourism.

Keywords: rural development; quality of life; migration balance; Eastern Moravia; cross-border projects

Citation: Vaishar, A.; Šťastná, M.; Kramáreková, H. Moravian–Slovak Borderland: Possibilities for Rural Development. *Sustainability* **2022**, *14*, 3381. https://doi.org/10.3390/su14063381

Academic Editor: John McDonagh

Received: 15 January 2022
Accepted: 11 March 2022
Published: 14 March 2022

1. Introduction

Over the last 30 years, several European states have collapsed: the USSR, Yugoslavia, and Czechoslovakia. It seems to be a good time now to find out how regional development is formed on the newly created border, which has suddenly become a harder or softer barrier, with the territory close to it becoming a periphery.

Our attention is focused on the Czech–Slovak border, where the development was relatively calm and which after a certain time became the internal border between EU states. Nevertheless, we believe that the question of the impact of the restoration of the state border here plays a role, even though the border is freely passable for people and goods and the language barrier is negligible. Legislative systems, social security systems and, finally, monetary systems have been separated. Relations with Slovakia have formally reached the level of relations with other EU states.

Of course, the situation is complicated. Other factors that are only marginally or not at all related to the new border can also play a role. For example, Eastern Moravia has become the most remote part of the Czech Republic in terms of the West–East gradient. Toušek and Tonev [1] showed that the interest of foreign capital in supporting the economy in districts in the Czech–Slovak borderland was by far the lowest in comparison with the borderland with all other neighbors, before joining the European Union. The structural reconstruction of the economy in market conditions certainly has its specifics. Everything takes place against the background of general trends in the transition to a post-industrial society, globalization, climate change and other current processes.

In our study, we raise the question of whether rural peripheral micro-regions (the term microregion is used in our work to distinguish it from administrative regions (NUTS 3); micro-regions, on the other hand, are real functional regions integrated in particular by commuting to work and services) on the eastern border of the Czech Republic show increased peripheralization after the division of Czechoslovakia and whether there are conditions and ideas to overcome them. The importance of this contribution lies in the fact that it draws attention to the border regions at the new state borders, which are dealing with the problem of division. Such studies dealing with the new borders between Russia and the former federal republics of the USSR or between the states of the former Yugoslavia are either missing or not part of the international literature. Moreover, in our case, there were multiple changes in the geopolitical situation over a relatively short period.

2. Theoretical Background

The key theoretical question is how to understand the concept of regional development and how it can be evaluated. Some authors identify development with quantitative growth [2]. This approach corresponds to the productive (modern, Fordist) period, and today it is more typical for developing countries [3], or for some post-Soviet approaches, which put economic factors of development in the forefront [4]. At the same time, Farooq and Ahmad [5] admitted that economic growth does not necessarily mean reducing poverty. Leick and Lang [6] argued that times of permanent growth of economy and population have come to an end for many European regions. It shows that it is necessary to change planning approaches for regions beyond growth.

Michalska-Żyła and Marks-Krzyszkowska [7] assumed that economic development and the development of quality of life cannot be equated. The policy of the European Union—although formally forcing multifunctional rural development—is focused primarily on the support of agriculture and tourism [8], which more or less corresponds to the traditional focus. It is to be feared that the more modern conception is hindered mainly by the lobbyist interests of agricultural entrepreneurs. Hrabák and Konečný [9] emphasize the importance of multifunctional agriculture for rural development, especially in mountain areas, which do not have favorable conditions for the development of traditional intensive agriculture.

In connection with the transition to a post-productive society, consumption in the broadest sense, i.e., the consumption of tangible and intangible values [10], including landscapes [11], buildings, traditions and habits, come to the fore. The quality of life or well-being of villagers (including entrepreneurs and visitors) is not determined by economic factors only [12]. Factors, such as access to education and medical care, personal safety, social inclusion, and participation in community life, but also subjective factors, such as satisfaction and happiness, come into play. The problem is that these factors are very difficult to identify and evaluate. However, if we are talking about regional development in rural areas, we need to focus on this direction because the possibilities of quantitative development encounter demographic, environmental and economic barriers. Many authors consider the quality of social capital to be a key factor in rural development at the local level [13].

Areas that are becoming borders can also become a periphery [14]. Peripheral regions are often characterized by depopulation, unemployment, lagging infrastructure, lack of investment and, ultimately, a deterioration in the quality of life of the local population. The border position may also mean the advantage of cooperation with the partner country, or even with a share of the grey economy [15]. Kühn [16] states that peripheralization is not just a function of distance, but that it is a multidimensional concept involving economic polarization, social inequality and the division of power. An important question is also related to the mental periphery, i.e., whether the inhabitants of remote regions perceive their location as peripheral and how the people from central regions perceive the periphery [17].

Depopulation is initially the result of a negative migration balance. Whereas young and educated people, in particular, are looking for a career in more developed areas and more attractive sectors, seniors remain in peripheral regions. This leads to secondary depopulation due to natural population decline [18]. This trend should not become irreversible and no longer solvable by regional policy instruments [19].

The approach and significance of border research have changed dramatically in the last decades [20]. The first expert analyses concerning state borders dealt with the delimitation of borders based on natural, historical, traffic, ethnic, religious and other characteristics of the territory. Their practical significance most often lay in the justification or rejection of territorial claims between neighboring states. State borders were usually more of a formal line for local people until the emergence of nation states, which began to defend national markets in the form of tariffs and other barriers. Often, state borders became a hard barrier and borders between enemy states on which fortifications and other military establishments were built. The extreme form was an iron curtain, or defended fences and walls, preventing the passage of people. Many changes took place in Central Europe, from the disintegration of Austria–Hungary through the disintegration of Czechoslovakia and Yugoslavia to the division and reunification of Germany [21]. Kolosov and Więckowski [22] pointed out several changes not only in the state borders, but also in their character, especially in Central and Eastern Europe in the 20th century.

Concerning the fall of the Iron Curtain, the interest of Czech geographers in border research has intensified. However, the greatest attention was paid to the border, which was occupied by the German Empire after the Munich Agreement and was where several problems accumulated. The expulsion of the majority of the German population after the Second World War and its replacement by Slavic immigrants is significant, which meant the interruption of centuries of development of localities and the arrival of inhabitants who had no previous experience with this type of territory [23]. Another problem is the sharp post-war decline in population, which, together with military measures at the state border, led to the demise of plenty of settlements. Other borders (except Slovakia) are also within easy reach of Prague and regional capitals, which has led to a significant increase in second homes. In contrast, the Czech–Slovak border was considered relatively problem free, was not burdened by ethnic conflicts, and had a more stable population for a long time. Perhaps that is why special attention was not given to this section of the borderland.

After experiencing two world wars, the western part of Europe concluded that it was necessary to change the nature of relations between states and, subsequently, change the state borders from lines of confrontation to areas of cooperation. These ideas were expressed in the creation of the European Union and later in the creation of the Schengen area, which allows the free movement of people and goods. The importance of research into state borders and border areas in Europe has increased since the fall of the Iron Curtain [24].

However, free movement is not enough. It is necessary to actively remove the burden of the past and to develop various forms of cooperation. That is why a system of Euroregions was created [25]. Euroregions are the EU's instrument for cross-border cooperation. Their purpose is, in fact, to overcome mistrust among the inhabitants of border regions and to build cultural and social bridges, creating a cross-border social capital [26]. The effectiveness of cross-border cooperation through Euroregions varies. In principle, the decisive factors are, inter alia, the different degrees of centralization in neighboring states, i.e., how far municipalities and other entities directly on the border are legally subjective to institutionalize cooperation. Equally important are motivations for cross-border co-workers. In some cases, it is simply a matter of overcoming the psychological barriers that were created in the past. The cooperation is directed from the cultural level—creating sub-state diplomacy [27], and in others, it may be the cooperation of a more complex level, including economic life [28]. The economic strength of neighboring regions is probably also important—whether these are economically developed or marginal regions, or how large the economic differences are between them. Medeiros [29] points out that EU cross-border cooperation programs are often used to finance the development of national border areas rather than to overcome cross-border barriers. However, the case of the Czech–Slovak border is different because it was a matter of dividing the state and thus maintaining an atmosphere of trust and cooperation.

Jeřábek et al. [30] distinguished three stages of cross-border cooperation of Czechoslovak and later Czech and Slovak regions: (a) the period 1989–1992, which is characterized by the so-called "wild" (spontaneous) cooperation without much coordination, mostly at the municipal level, (b) the period 1993–1996, when the European Union enters this cooperation, which sees (border) regions as the engine of cross-border cooperation (so-called Euroregions are created along the borders), and (c) the period after 1997, when this regional institutional cross-border cooperation coordinated by the European Union takes concrete form and when cooperation takes place with local institutions. Cross-border cooperation in the Moravian–Slovak borderland was supported by the European Union even before the accession of both countries to the EU [31]. At present, the Moravian–Slovak border is covered by the Euroregions Beskydy/Beskidy (with Poland as the third partner) in the northern part [32] and Bílé Karpaty/Biele Karpaty in the southern part of the borderland. The question of overcoming the psychological barriers of extraordinary hostility is out of the question in the case of Czech–Slovak relations. The regions on both sides are also ethnically and linguistically close because the ethnographic transition between Slovakia and Moravia is not sharp, but gradual.

There is not much work comparing rural development on the Moravian–Slovak border. Smékalová et al. [33] compared the effectiveness of the use of structural funds in the Zlín and Trenčín regions with ambiguous results. The regions on the Slovak side are among the most advanced in Slovakia in terms of GDP per capita. However, this does not have much effect on the development of cross-border cooperation. If there is cross-border commuting from Slovakia, it goes more to Austria [34].

The key question is what cooperation should border regions that are not relatively economically advanced be based on. One of the possibilities is the joint protection of nature and natural values. The second option is the development of cross-border tourism. On the other hand, economic cooperation and cross-border commuting to work are almost out of the question.

Of course, the changes in society as a whole cannot be neglected either. In connection with the transition to a post-industrial society, the European countryside is changing from an area for agricultural production to a multifunctional landscape [35,36]. Murdoch and Pratt [37] even discussed post-rural areas, whereas Oliva and Camarero [38] used the terms de-peasantization and de-agrarianization. These changes are accompanied by a shift in labor from the manufacturing sector to services. It is the speed of this process that can be a measure of the achieved level of regional development. At the same time, it seems clear that large cities are characterized by the fastest growth in the services sector, while the countryside is lagging. The reason is obvious. Most services tend to be concentrated, and therefore located in the centers of each hierarchical level. One of the services that partially deviates from this rule is tourism services that are tied to deconcentrated attractions —natural beauty or cultural heritage. This is important, especially in lagging areas [39].

The role of the development of cultural tourism in the post-industrial countryside is emphasized, for example, by Vidickiene, Vilke and Gedminaite-Raudone [40], who discussed transformative tourism. The development of cultural tourism in rural areas is made possible, inter alia, by the development of tourism from an elite sector to a sector for the general masses of Europe [41]. Băndoi et al. [42] pointed to the positive relationship between the development of tourism and the quality of life of local people—although this relationship may also be contradictory, especially in the case of over tourism.

The question remains of what forms of tourism are possible for remote rural microregions. By far, the most frequent is second housing in cottages and chalets. This form does not bring much financial benefit to rural regions, but has saved many rural buildings and, through contacts between cottagers and locals, enables contacts between urban and rural lifestyles and supports regional identity [43]. From a different perspective, it can also be preparation for permanent housing. In addition, the rental of individual cottages is developing today. Other possible forms of tourism in these areas are outdoor recreation (summer by the water and winter in mountains), ecotourism, nostalgic tourism and sightseeing cultural tourism. These forms of tourism do not bring much financial benefit—unless they are of a mass nature. However, building mass tourism is not in line with the quality of life of the locals. Efforts to build capacity for more affluent clients have not been met with success [44]. Agritourism is not yet very developed in Czech remote rural areas [45].

The paper then focuses on the evaluation of the possibilities of development in rural areas of Eastern Moravia in the near future. The main research question is the possibility of developing regions that have become remote as a result of the changing geopolitical situation.

3. Methods

The historical–geographical method and the path-dependency method were used to understand the current state of the studied region. This method is based on the experience that past developments affect the current situation [46]. It can be about using past experience or avoiding past mistakes.

The basic methodological problem is to find indicators and criteria for evaluating rural development in terms of the quality of life of its inhabitants. Of course, sociological methods of a questionnaire survey, guided interviews and other procedures are offered. These methods—if they are to be sufficiently meaningful—require complex preparation, ensuring representativeness, careful implementation and evaluation [47]. Ultimately, they are costly and time consuming.

Another possibility is to use a set of variously selected economic, social, demographic, environmental and political indicators with a greater or lesser relationship to quality of life. Kebza [48], in analyzing the degree of peripherality, used six indicators: unemployment rate, net migration rate, age dependency index, gross leasable area, number of students and number of occupied job opportunities. The main problem is to find the weights of individual indicators in the whole system and sometimes to determine the relationship of individual indicators to quality of life. The question of the subject's quality of life also arises here, as seniors will probably have different preferences from juniors and employees will have different preferences from entrepreneurs. Therefore, this approach is more suitable for the evaluation of individual aspects of quality of life and for smaller areas where field experience can be applied.

Would it be possible to find any publicly available and frequently updated statistical indicators that would indicate the objective and subjective aspects of the qualitative aspects of rural development? We believe that the migration balance speaks very comprehensively about the quality of life. We assume that contemporary people in Europe are moving from places with a lower quality of life to places with a higher quality of life—no matter how individuals define the quality of life. Hoogerbrugge and Burger [49] admitted that at least part of selective migration can be explained by differences in life satisfaction—especially on the side of urban–rural migration. Of course, it can be argued that some people cannot move out of low-quality-of-life areas for financial or other reasons. Therefore, it seems appropriate to supplement this indicator with the number of newly built dwellings. Although in this case, other factors, especially the price and availability of building plots, may play a role, we do not assume that anyone would build an apartment in an area they do not like.

The migration balance (the number of immigrants to individual settlements in the area minus the number of emigrants from them) and completed dwellings is calculated for the five years of 2016–2020. The studied area is monitored according to the catchment areas of authorized municipal authorities to capture even smaller intra-regional differences. Another additional indicator is the level of unemployment (number of available job seekers to the number of inhabitants aged 15–64), which may be one of the motives for migration—even though this indicator is losing its significance compared to the past, due to an increasing proportion of the population being pensioners. In Czech conditions, unemployment is not a general problem, but its locally increased level may signal local disparities in the labor market. As an additional indicator, the age structure of migrants was analyzed according to their age groups and the main directions of migration from the region of interest.

4. Study Area

The studied area (Figure 1) was delimited by catchment areas of municipalities with extended powers on the border with the Slovak Republic. (It is an intermediate stage between municipalities (LAU 2) and districts (LAU 1). Municipalities with extended powers (usually small- or medium-sized cities) carry out some professional agendas for municipalities in their background, which are not effective or possible to perform in too-small municipalities. However, municipalities with extended powers are not superior to municipalities in their catchment area. In some cases, there is another intermediate stage—municipalities with an authorized municipal office, if the catchment areas of municipalities with extended powers are too large and it is expedient to carry out some agendas closer to the population.) These municipalities are located in three NUTS 3 regions: South Moravian (Hodonín and Veselí nad Moravou), Zlín (Uherský Brod, Luhačovice, Valašské Klobouky, Vsetín, Rožnov pod Radhoštěm) and Moravian-Silesian (Frýdlant nad Ostravicí, Jablunkov (the catchment area of Frýdek-Místek, which extends only marginally to the border, was omitted because its center has more than 50,000 inhabitants and a population density that reaches 232 people per km^2; for this reason, it can be hardly included among the peripheral rural regions)). Against them, there are nine administrative districts of three regions on the Slovak side: Žilina, Trenčín and Trnava. The area under study is mostly covered by the mountain ranges of the White Carpathians and Moravian-Silesian Beskid Mountains and their foothills and side valleys. The only exception is the microregion of Hodonín, which already falls into the Lower Moravian lowland and wine-growing area. The rest of the area is destined for extensive organic farming, which, however, creates its high dependence on subsidies. The total area is 2934 km^2. In the Slovak part, the territory borders with Kysucké Beskydy Mts., Javorníky (Maple Mts.), White Carpathian Mts. and Záhorie lowland.

Figure 1. The area under study. Drawn by J. Pokorná.

The area was covered mainly by forests (46.8%) at the end of 2019. Arable land accounted for 21.4%, permanent grassland 18.3%. Other types of land use were insignificant: permanent crops (vineyards, gardens, orchards 3.4%, water areas 1.4%, and built-up areas 1.5%). At the end of 2020, almost 340,000 inhabitants lived in the studied area, which, due to the mountainous nature of the relief, represents an unexpectedly high density of 116 inhabitants/km^2. There are a total of 169 municipalities in the area, of which only 8 have less than 200 inhabitants. There are 72 large rural municipalities with more than a thousand inhabitants and 48 medium-sized rural municipalities with 500 to 999 inhabitants. There are 13 towns with more than 5 thousand inhabitants, of which over 20 thousand have Vsetín (26 thousand) and Hodonín (24 thousand). This structure is also non-standard in the Czech conditions, where very small and small rural municipalities predominate. Not all municipalities are compact villages. In some places, there is a scattered settlement.

The population of the studied area is 45.4% employed in manufacturing industries (agriculture, forestry, fishery, manufacturing and building industry). (The data are from the 2011 census, as data from the 2021 census have not yet been published. It can be assumed that the ratios are maintained, although the shares of employees in the manufacturing sectors are generally declining). This is 10.5 percentage points more than the national average (see Table 1). From this point of view, Eastern Moravia is a significantly rural and lagging-behind area. This corresponds to the share of people over the age of 15 with a high school diploma, which reaches 39.2%, which is 4.4 percentage points less than the national average. Only the districts of Luhačovice (with a spa function) and Frýdlant nad Ostravicí (the goal of suburbanization of Ostrava and its agglomeration) have industrial employment at the national average.

Table 1. Shares of economically active people employed in individual sectors of the economy (2011), age index (2020), unemployment rate (December 2021).

Microregion	Primary Sector	Secondary Sector	Tertiary Sector	Age Index	Unemployment Rate
Hodonín	3.4	38.4	58.2	1.41	5.5
Veselí nad Moravou	4.4	43.7	51.9	1.62	5.2
Uherský Brod	3.2	46.2	50.6	1.45	2.2
Luhačovice	2.5	36.6	60.9	1.47	2.0
Valašské Klobouky	3.8	45.3	50.9	1.22	1.8
Vsetín	3.0	42.7	54.3	1.34	3.7
Rožnov pod Radhoštěm	2.2	45.9	51.9	1.42	2.9
Frýdlant nad Ostravicí	3.3	36.1	60.6	1.31	3.3
Jablunkov	2.5	41.8	55.7	1.06	2.7

Data source: Czech Statistical Office Prague, Ministry of Labor and Social Affairs. Own elaboration.

At the end of 2020, the population age index (the share of the number of seniors over the age of 50 and children aged 0–14) in Eastern Moravia was 1.35 in comparison with the national value of 1.26. The population of the monitored region is, therefore, relatively older (see Table 1). The oldest population has the district of Veselí nad Moravou in the southern part of the border, while the northernmost district of Jablunkov has a population relatively young, even compared to the national average.

According to the Ministry of Labor and Social Affairs of the Czech Republic, unemployment in the region was 3.6% in December 2021, while national unemployment was 3.5% (see Table 1). At present, the problem is the availability of a skilled workforce to maintain production. Increased unemployment can be found in both of the southernmost micro-regions, while in the middle of eastern Moravia, the value of this indicator is around 2%, which signals a shortage of the labor force. On the Slovak side, unemployment is similar to the top in the northern section of the borderland, as the districts of Western Slovakia have the lowest level of unemployment in the country: Trnava region 4.2%, Trenčín region 4.3%, and Žilina region 5.3%.

The area includes several specific ethnographic areas with a distinctive culture. From the south, these are Horňácko (as part of Moravian Slovakia), Wallachia, Lachia and Těšín

Silesia (in the last two, elements of Polish culture can also be found). This, on the other hand, may signal a certain conservatism of the local population. The area has many attractions for the development of cultural tourism. Natural values are protected within the protected landscape areas of the White Carpathians and the Beskid Mountains, which have their counterparts in protected landscape areas Biele Karpaty and Kysuce in Slovakia and occupy a significant part of the territory. Conditions allow summer and winter tourism. In the territory of Eastern Moravia, there are also many monuments of archaeological nature, medieval fortifications, church and secular buildings, important open-air museums in Rožnov pod Radhoštěm and Strážnice, work and native places of important personalities, such as J.A. Komenský, L. Janáček and others, as well as modern attractions. The local, partly lively folklore, manifested in folk art, customs and traditions, is unique. The most important Moravian spa Luhačovice (Figure 2) is also located here.

Figure 2. The attractive architecture of the best-known Moravian spa Luhačovice. Source: the authors.

There are 498 collective accommodation facilities in the region (2020), which have 9216 rooms with 27,219 beds. This represents 9.3 beds per km^2. This value slightly exceeds the national average, which is 7.5 beds per km^2. In the last pre-COVID year 2019, 2,315,761 overnight stays were recorded, of which 35.6% were in the Luhačovice area. It represents 6.8 overnight stays per inhabitant (to compare with the national average of 5.3 overnight stays per inhabitant). Of the total overnight stays, 10.8% were foreigners (in the national average it was 47.7%), so the region is oriented to domestic tourism. To the accommodation capacity, it is necessary to add 4524 unoccupied flats used for recreation (2011). In addition, Kubeš [50] estimates the number of holiday cottages in the Beskydy region at 11,200. This can mean around 60,000 beds—double the number of beds in collective facilities. However, these beds are mostly used only for individual recreation of the owners and their friends.

5. Short Historical Overview

There is an opinion that the Czech–Slovak border is a new one. However, this only applies in the short term. In fact, both sides of the border were part of one state unit during the Great Moravia (The exact name of the then state is not known. The term Great Moravia is of a later date and serves primarily to separate today's Moravia from the then empire, which occupied a relatively large part of Central Europe, including Bohemia, Lusatia, Silesia, Pannonia and other territories. There are also uncertainties about the territorial area and centers of Great Moravia [51].), which originated from the connection of Moravian and Nitra principalities in 833 and ended probably by the refutation by Avars in 907. After that, the border was the division between the Moravian Margraviate/Czech Kingdom and Hungary for 1011 years. Although for some period the Czech lands and Hungary were personally united by the same ruler, the border has always existed here.

It should be noted that, except for the southernmost part, it was a very sparsely populated and late colonized area. Wallachian colonization did not come here until the 16th and 17th centuries. The population subsisted on pastoralism. Family breadwinners often went throughout the monarchy and beyond during the summer seasons. Many people also emigrated overseas.

Pastoralism has led to extensive deforestation. The situation changed during industrialization, which released the pressure on the landscape. Changes in agriculture in the second half of the 20th century—collectivization (the 1950s), mechanization (1960s), industrialization (1970s) and intensification (1980s) resulted in an increase in arable land in the foothills and mountain areas [52]. In addition to compact villages, the settlement also contains small settlements or solitudes. The northern part of the borderland has been under the influence of the Ostrava Industrial Agglomeration since the 19th century, based on hard coal mining and heavy industry.

The Czechoslovak period, although relatively short, brought very significant changes to the territory of Eastern Moravia. The shoe company Bat'a was developed in Zlín. It was a complex company, which operated not only in footwear, but in many other industries (manufacture of tires, technical rubber, man-made fibers, toys, metalworking machines, knitting machines, aircraft, and bicycles). The group employed 67,000 people. The company introduced a special motivation system [53]. To transport coal from the South Moravian lignite district, a canal was built. Branch factories were built in borderland small towns. This way of conducting business significantly affected the entire business environment of the Zlín and Vsetín areas.

Another impetus for the industrialization of the region was the approaching second world war, which evoked a boom of the Czechoslovak armament industry [54]. The Czech lands were surrounded by Hitler's Germany on three sides. All large Czech armories were directly threatened by bombing by German planes taking off from Austria and landing in Silesia (or vice versa). Therefore, each Czech armory built one or two daughter armories in mountain valleys in Eastern Moravia or Western Slovakia, where these factories were more sheltered from enemy air raids, and where the Czechoslovak army would withdraw from enemy attacks. The result was a series of armaments and ammunition companies in Uherský Brod, Bojkovice, Slavičín, Vsetín, and Kunovice. These companies have significantly contributed to the industrialization of the former agricultural and forestry areas.

The connection of Czechia and Slovakia required the construction of large-capacity communications in the west–east direction. Originally, the only long-distance connection between Czechia and Eastern Slovakia was the Košice-Bohumín Railway. Additional railways were put into operation through the Lyska and Vlára passes. More road connections have been built.

The flat and more accessible southern part of the Moravian-Slovak border (the Hodonín area) was surprisingly industrialized later; perhaps because it provided favorable conditions for intensive agriculture and viticulture. The sugar industry was the first leading branch here. In the 1950s, a power plant processing local lignite and a modern sugar factory was added to the 19th-century tobacco factory. Quality oil is mined in the area.

In any case, at the end of the Czechoslovak period, East Moravia, formerly a pastoralist and handicraft region, was an industrialized area, including the relevant consequences for the qualification structure of the population, the infrastructure and the like.

The demise of Czechoslovakia in 1993 changed the geopolitical position of the region. From the area in the middle of the state, it became a peripheral territory—moreover, on the least attractive eastern edge. Complications began to manifest gradually. The state border remained freely permeable for citizens of the Czech Republic and the Slovak Republic, but not for citizens of third countries. This limited tourism because foreign tourists could not use hiking and skiing trails passing from one state to another. Another problem was the separation of currencies. Social security systems (pensions) have also been separated, so it has become disadvantageous for the citizens of Czechia to work in Slovakia. It is obvious that the intensity of cross-border traffic has decreased significantly [55] as well as cross-border commuting [56].

The current integration of Czechia and Slovakia first into the European Union in 2004 and later into the Schengen area from its inception in 2007 brought new impetus. Theoretically, the state border should change in the line of cooperation; however, some barriers remain at the local level. This is both a natural barrier (mountains and a watercourse) and a psychological barrier [57], which in this case is less significant.

6. Results

Between 2016 and 2020, 19,792 people immigrated to the region, while 21,217 inhabitants emigrated (Table 2). All districts recorded a negative migration balance, except for Frýdlant nad Ostravicí, which is highly positive (+30.5‰). The Frýdlant region lies in the hinterland of the Ostrava industrial agglomeration and is therefore the subject of the remote suburbanization of Ostrava. On the other hand, the districts of Luhačovice (−12.3‰), Valašské Klobouky (−11.6‰) and Veselí nad Moravou (−11.0‰) have a relatively high negative balance of migration. These are districts of smaller centers. The micro-regions of the large centers of Vsetín, Hodonín, Uherský Brod and Rožnov pod Radhoštěm have high absolute population declines (over 500 persons), but the emigration shares are relatively lower.

Table 2. Migration and new flats index in the period 2016–2020.

Microregion	Population	Immigrants	Emigrants	Balance	New Flats per 1000 Inhabitants
Hodonín	60,579	3457	3738	−4.6‰	8.8
Veselí nad Moravou	37,498	2027	2439	−11.0‰	9.2
Uherský Brod	51,960	2806	2973	−3.2‰	11.4
Luhačovice	18,567	1337	1565	−12.3‰	9.6
Valašské Klobouky	23,053	1140	1411	−11.6‰	10.5
Vsetín	65,126	2710	3270	−8.6‰	10.8
Rožnov pod Radhoštěm	35,043	2061	2130	−2.0‰	15.8
Frýdlant nad Ostravicí	25,135	2974	2208	+30.5‰	15.2
Jablunkov	22,676	1459	1483	−1.1‰	14.1

Data source: Czech Statistical Office Prague. Own elaboration.

The position of the largest towns in the region is interesting. All four with a population over 15,000 (Hodonín, Vsetín, Rožnov pod Radhoštěm and Uherský Brod) are significantly passive in terms of migration. They always account for the majority of the negative migration balance of their micro-regions. It can be said that without these cities, the migration balance of their micro-regions would be almost balanced. It is also possible that many people are moving from these centers to the villages around them, and therefore it is a kind of micro-suburbanization.

An interesting fact is that the negative migration balances had a decreasing tendency during the monitored five years (Figure 3). In 2020, the region as a whole recorded even a very slight migration increase. This is to some extent in line with the reversing trend from rural-to-urban to urban-to-rural migration.

Of the total number of 3025 emigrants outside the studied region in 2020, people in the age group 0–14 years made up 22.6%, people in the productive age group 15–64 years 71.7% and seniors aged 65 and over only 5.7%. The age structure of immigrants is similar. Of the 2787 immigrants, 24.8% were children and young people, 69.1% were people of working age and 6.1% were seniors. Therefore, it is not obvious that young people are moving out and being replaced by seniors.

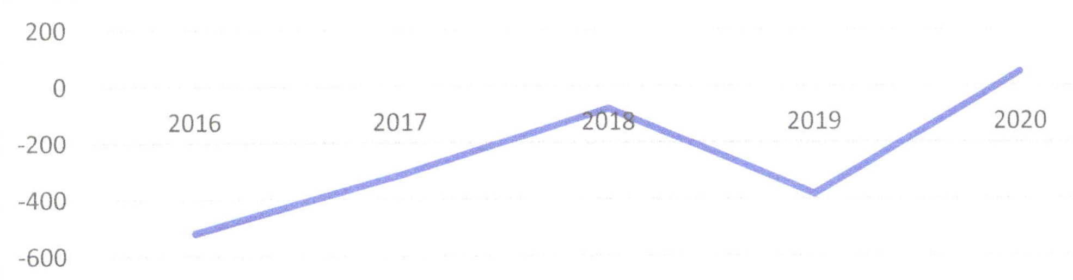

Figure 3. Development of the migration balance in the region under study 2016–2020. Data: Czech Statistical Office Prague. Own elaboration.

Of the total number of emigrants, 11.7% went to Prague and 9.4% to Brno. Ostrava is more important for the Frýdlant nad Ostravicí microregion. Significant migratory flows are directed to nearby medium-sized cities. For the southern part of Eastern Moravia, a center is Uherské Hradiště, for the central part, Valašské Meziříčí and, to a limited extent, Zlín, and for the northern part, Frýdek-Místek and Třinec. Certain numbers are directed to neighboring micro-regions and the rest is dispersed. Immigration flows are the strongest from Ostrava and its agglomeration to the micro-regions of Frýdlant nad Ostravicí and partly Jablunkov.

As a result, current migration is in line with traditional patterns. While the migration of seniors is minimal, mainly people in the first half of the productive age with children migrate. In addition, only about a quarter of emigrants go to the big cities. The rest are looking for employment in nearby medium-sized cities, corresponding to their qualification structure. It may indicate that they do not intend to sever ties with their original residence.

New apartments are appearing more in the northern part of the territory. Surprisingly, however, relatively, most of the new dwellings per 1000 inhabitants were built not in the Frýdlant district, but the Rožnov pod Radhoštěm district, followed by Frýdlant nad Ostravicí (Figure 4) and Jabunkov. This means that high immigration into the Frýdlant microregion is not accompanied by adequate housing construction. Immigrants, therefore, partly use the existing housing stock. On the contrary, the lowest numbers of new dwellings per thousand inhabitants are in the southern districts of Hodonín, Veselí nad Moravou and Luhačovice.

Figure 4. Frýdlant nad Ostravicí: The new market street in a small town. Source: authors.

The level of unemployment in June 2021 fell below 3% in the micro-regions Uherský Brod, Luhačovice, Valašské Klobouky Rožnov pod Radhoštěm and Jablunkov, i.e., mainly in the central and northern parts of the Moravian-Slovak borderland. The highest unemployment can be found in the southern micro-regions of Hodonín and Veselí nad Moravou.

7. Discussion

Data on migration as well as data on newly built dwellings are published annually by the Czech Statistical Office at the municipal level. This means that the analysis can be performed and repeated for any territory composed of municipalities, annually and without additional financial costs.

In terms of practical use, there is one significant objection to the methodology used. There is a relatively long delay between a change in the quality of life and a migration decision. Therefore, the methodology reflects rather the past characteristics of the region. Although it can be assumed that quality of life does not change too quickly regionally, in the event of significant changes in the geopolitical, economic or other situation, other indicators must also be used. However, the proposed indicator enables long-term monitoring of trends after individual years, so it is possible to respond to emerging problems relatively quickly. The second objection is that the proposed methodology does not address the differentiated values and demands of different demographic, cultural and social groups of the population. While mostly career-oriented people leave peripheral rural regions (permanently or temporarily), environmentally oriented people can stay or immigrate.

The uniqueness of the Frýdlant nad Ostravicí district shows that the quality-of-life evaluation based on migration cannot be assessed absolutely but always as differences between the sources and targets of migration. The differences between the comprehensively understood quality of life in Ostrava (respectively its image) and the Frýdlant nad Ostravicí district are so significant that they cause relatively intense migratory movements. Other parts of Eastern Moravia do not have such counterparts.

Although there are plenty of job opportunities in the industry, except for the southernmost part of the study area (the region even seems to be facing labor shortages), structural restructuring is inevitable. The high share of industry is no longer a sign of progressiveness, but rather of backwardness. Job opportunities in the industry will decline in the long run due to changes in labor productivity and the overall post-productive transformation. The transfer of a significant part of the workforce from industry to services is also important for cultivating the human factor within the post-productive economy.

A certain positive signal is a fact that the perception of Eastern Moravia is not explicitly peripheral. From the center's point of view, the border periphery from which the German population was evicted after World War II is far stronger and therefore has to deal with the consequences of ethnically conditioned population exchange, disruption of historical continuity and difficult relationship-building. East Moravia does not have these problems. From the point of view of its inhabitants, these are traditional ethnographic regions connected by a distinctive culture and a high degree of regional identity.

To evaluate the overall situation, we summarized the strengths and weaknesses of the region, its opportunities and threats (Table 3).

While the first structural change from a pastoral and forestry area to a manufacturing region was the result of a change from a pre-productive to a productive society, the forthcoming change from a production region to a consumer region is a consequence of the transition to a post-industrial society. Coincidentally, both transitions were accompanied by changes in the geopolitical position of the region. A structure consisting of organic multifunctional agriculture, innovative processing industries and domestic tourism should be created. As far as tourism is concerned, it should focus on combining different forms with an emphasis on enhancing experience and quality, but not exclusivity. The purpose is not to create jobs, but to change the way of life and the system of values of the local population.

Despite the establishment of Euroregions and the removal of some barriers, a significant economic impulse cannot be expected from cooperation with the Slovak side. At present, the development of Slovak–Czech cross-border cooperation is ensured mainly by the objectives of the INTERREG V-A Slovak Republic—Czech Republic Cooperation Program—which in the 2014–2020 programming period focused on four priority axes: exploitation of innovation potential, quality environment, development of local initiatives and technical assistance. The SK-CZ Regional Advisory Center Project plays an important role in all six regions on the Slovak–Czech border. Specific cooperation is mainly focused on the development of cross-border cultural tourism. These include the Cyril and Methodius Trail, support for the celebrations of the fraternity of Czechs and Slovaks at Velká Javorina Hill. Cycling routes and nature trails are being built, and the historical heritage in the border area is being restored. There are joint nature conservation projects, such as combating erosion and drying up wetlands. Cross-border cooperation is focused on tourism, cultural heritage and nature conservation corresponding to the geographical character of the studied area.

Table 3. Strengths, weaknesses, opportunities and threats. Source: own elaboration.

Strengths	**Weaknesses**
• Valuable attractions for the development of domestic tourism • Population, firmly rooted in regional and local identity	• A remote location from European and national points of view • Too much grounding in the manufacturing industry
Opportunities	**Threats**
• Possible use of subsidies for primary activities and tourism • A growing interest of residents in domestic tourism	• Decline or loss of competitiveness of the engineering and electrical engineering industries

Of course, the results of this analysis could be further dissected. The situation may be different in different regions, and it would be possible to go into a more detailed view at the level of authorized municipal authorities. Candidates for a deeper analysis would be, for example, the micro-regions of Velká nad Veličkou, Slavičín, Bojkovice (Figure 5), Brumov-Bylnice, and Karolinka, that is, micro-regions located directly on the border, which have indistinct centers. There would be scope for the use of other methods, such as qualitative sociological research methods, for example.

Figure 5. Bojkovice. The church indicates a more famous history than the present. Source: the authors.

8. Conclusions

If we are to answer the question of whether Eastern Moravia shows signs of negative trends in rural development after the change in the geopolitical situation, we must answer in the affirmative. A negative migration balance is the only exception. However, the overall situation is regionally differentiated. While the northern part represents the rural hinterland of the Ostrava agglomeration, the southern part shows the least favorable characteristics in all respects. However, the situation does not seem to be deteriorating. Large villages and small towns are still stable, providing a sufficient local market.

However, East Moravia is located on the threshold of the necessary structural reconstruction from a region focused on the manufacturing industry to a region with an increased share of services. The opportunity is the further development of tourism services with a focus on soft and cognitive forms, including cultural tourism. Although the current capacity of accommodation facilities is above average, reserves could certainly be found. Due to the remoteness, foreign tourists probably come mainly from neighboring countries—Slovakia and Poland.

Further research should focus in particular on the evolution of the situation over time, given that the situation is still evolving. Migration trends are beginning to turn toward urban-to-rural migration. However, the development is selective and may not be permanent in individual cases. Another direction of research could be to compare different rural regions, even in an international framework, as the criteria used are so simple that there should be no fundamental differences in the databases. The third direction of research could be detailed in selected micro-regions, which would allow the application of other methods. The study of spatial aspects of drawing EU financial resources at the local level can also provide interesting information about the absorption capacity of this area and the diversification of the project content [58].

In any case, border changes, which are usually agreed upon in capitals or other major centers, have consequences in places at the state border [59]. These changes should be of interest to geographers and other professionals, as they can significantly affect the quality of life of border residents.

Author Contributions: Conceptualization, A.V. and M.Š.; methodology, A.V.; investigation, A.V., M.Š. and H.K.; writing—original draft preparation, A.V.; writing—review and editing, M.Š. and H.K.; funding acquisition, M.Š. All authors have read and agreed to the published version of the manuscript.

Funding: This research was funded by the EU HORIZON 2020 project Social and Innovative Platform On Cultural Tourism and its Potential Towards Deepening Europeanisation (https://cordis.europa.eu/project/id/870644 (assessed on 14 January 2022)), ID 870644, funding scheme Research and Innovation action, call H2020-SC6-TRANSFORMATIONS-2019.

Institutional Review Board Statement: The study was conducted according to the guidelines of the Declaration of Helsinki and approved by the Ethics Committee of Mendel University in Brno (protocol code No. 13, date of approval: 8 October 2020).

Informed Consent Statement: Not applicable.

Data Availability Statement: Not applicable.

Conflicts of Interest: The authors declare no conflict of interest. The funders had no role in the design of the study; in the collection, analyses, or interpretation of data; in the writing of the manuscript; or in the decision to publish the results.

References

1. Toušek, V.; Tonev, P. Foreign direct investment in the Czech Republic (with the emphasis on border regions). *Acta Univ. Carol. Geogr.* **2003**, *38*, 445–457.
2. Wu, J.; Zhuo, S.; Wu, Z. National innovation system, social entrepreneurship, and rural economic growth in China. *Technol. Forecast. Soc. Chang.* **2017**, *121*, 238–250. [CrossRef]
3. Sertoğlu, K.; Ugura, S.; Bekun, F.V. The contribution of agricultural sector on economic growth of Nigeria. *Int. J. Econ. Financ. Issues* **2017**, *7*, 547–552.

4. Petre, I.; Ion, R. The impacts of the investments in agriculture on economic growth in rural communities in Romania. *Ekon. Poljopr.* **2019**, *66*, 955–963. [CrossRef]
5. Farooq, S.; Ahmad, U. Economic Growth and Rural Poverty in Pakistan: A Panel Dataset Analysis. *Eur. J. Dev. Res.* **2020**, *32*, 1128–1150. [CrossRef]
6. Leick, B.; Lang, T. Re-thinking non-core regions: Planning strategies and practices beyond growth. *Eur. Plan. Stud.* **2018**, *26*, 213–228. [CrossRef]
7. Michalska-Żyła, A.; Marks-Krzyszkowska, M. Quality of Life and Quality of Living in Rural Communes in Poland. *Eur. Countrys.* **2018**, *10*, 280–299. [CrossRef]
8. Kovács, Z.; Csachová, S.; Ferenc, M.; Hruška, V. Development policies on rural peripheral areas in Visegrád countries: A comparative analysis. *Stud. Obsz. Wiej.* **2015**, *39*, 77–102. [CrossRef]
9. Hrabák, J.; Konečný, O. Multifunctional agriculture as an integral part of rural development: Spatial concentration and distribution in Czechia. *Nor. Geogr. Tidsskr.* **2018**, *72*, 257–272. [CrossRef]
10. Lekakis, S.; Dragouni, M. Heritage in the making: Rural heritage and its meiosis at Naxos Island, Greece. *J. Rural Stud.* **2020**, *77*, 84–92. [CrossRef]
11. Di Fazio, S.; Modica, G. Historic Rural Landscapes: Sustainable Planning Strategies and Action Criteria. The Italian Experience in the Global and European Context. *Sustainability* **2018**, *10*, 3834. [CrossRef]
12. Vaishar, A.; Vidovićová, L.; Figueiredo, E. Quality of Rural Life. Editorial. *Eur. Countrys.* **2018**, *10*, 180–190. [CrossRef]
13. Permingeat, M.; Vaneste, D. Social capital in rural development projects in Europe–Three LEADER cases in Walonia analysed. *Belgeo* **2019**, *20*, 34979. [CrossRef]
14. Poeta Fernandes, G. Rural depopulation, social resilience and context costs in the border municipalities of central Portugal. Dichotomies of social reorganization vs absence of public policies. *Econ. Agrar. Recur. Nat.* **2019**, *19*, 121–149. [CrossRef]
15. Stryjakiewicz, T. The changing role of border zones in the transforming economies of East Central Europe: The case of Poland. *GeoJournal* **1998**, *44*, 203–213. [CrossRef]
16. Kühn, M. Peripheralization: Theoretical Concepts Explaining Socio-Spatial Inequalities. *Eur. Plan. Stud.* **2015**, *23*, 367–378. [CrossRef]
17. Willett, J.; Lang, T. Peripheralisation: A Politics of Place, Affect, Perception and Representation. *Sociol. Rural.* **2018**, *58*, 258–275. [CrossRef]
18. Johnson, K.M.; Lichter, D.T. Rural depopulation: Growth and decline processes over the past century. *Rural Sociol.* **2019**, *84*, 3–27. [CrossRef]
19. Pinilla, V.; Sáez, L.A. What Do Public Policies Teach Us About Rural Depopulation: The Case Study of Spain. *Eur. Countrys.* **2021**, *13*, 330–351. [CrossRef]
20. Anderson, A.; O'Dowd, L. Borders, Border Regions and Territoriality: Contradictory Meanings, Changing Significance. *Reg. Stud.* **1999**, *33*, 593–604. [CrossRef]
21. Havlíček, T.; Jeřábek, M.; Dokoupil, J. *Borders in Central Europe after the Schengen Agreement*; Springer Nature: Cham, Switzerland, 2018.
22. Kolosov, V.; Więckowski, M. Border changes in Central and Eastern Europe. An introduction. *Geogr. Pol.* **2018**, *91*, 5–16. [CrossRef]
23. Vaishar, A.; Dvořák, P.; Nosková, H.; Zapletalová, J. Present consequences of post-war migration in the Czech countryside for regional development. *Quaest. Geogr.* **2017**, *36*, 5–15. [CrossRef]
24. Paasi, A. A border theory: An unattainable dream or a realistic aim for border scholars. In *The Routledge Research Companion on Border Studies*; Wastl-Walter, D., Ed.; Routledge: London, UK, 2011.
25. Noferini, A.; Berzi, M.; Camonita, F.; Durà, A. Cross-border cooperation in the EU: Euroregions amid multilevel governance and re-territorialization. *Eur. Plan. Stud.* **2020**, *28*, 35–56. [CrossRef]
26. Böhm, H.; Opioła, W.; Drosik, A. Cross-border social capital. An analysis of selected elements as exemplified by the Praded Euroregion. In *Old Borders–New Challenges, New Borders–Old Challenges*; Janczak, J., Ed.; Logos Verlag: Berlin, Germany, 2018; pp. 99–113.
27. Klatt, M.; Wassenberg, B. Secondary foreign policy: Can local and regional cross-border cooperation function as a tool for peace-building and reconciliation. *Reg. Fed. Stud.* **2017**, *27*, 205–218. [CrossRef]
28. Pupier, P. Spatial evolution of cross-border regions. Contrasted case studies in North-West Europe. *Eur. Plan. Stud.* **2020**, *28*, 81–104. [CrossRef]
29. Medeiros, E. Should EU cross-border cooperation programmes focus mainly on reducing border obstacles. *Doc. Anàl. Geogr.* **2018**, *64*, 467–491. [CrossRef]
30. Jeřábek, M.; Dokoupil, J.; Havlíček, T. *České Pohraničí–Bariéra Nebo Prostor Zprostředkování*; Academia Praha: Prague, Czech Republic, 2004.
31. Rajčáková, E.; Švecová, A. Cross-Border Cooperation in Slovak-Czech Border Region under EU Programmes. *Eur. Countrys.* **2013**, *5*, 133–145. [CrossRef]
32. Howaniec, H.; Lis, M. Euroregions and Local and Regional Development—Local Perceptions of Cross-Border Cooperation and Euroregions Based on the Euroregion Beskydy. *Sustainability* **2020**, *12*, 7834. [CrossRef]
33. Smékalová, L.; Hájek, O.; Kubík, J.; Škarka, M. Management of European projects by Czech and Slovak entrepreneurs in Bíle-Biele Karpaty Euroregion. *Econ. Sociol.* **2016**, *9*, 69–85. [CrossRef]

34. Cavallaro, F.; Dianin, A. Cross-border commuting in Central Europe: Features, trends and policies. *Transp. Policy* **2019**, *78*, 86–104. [CrossRef]
35. Ling, C.; Handley, J.; Rodwell, J. Restructuring the post-industrial landscape: A multifunctional approach. *Landsc. Res.* **2007**, *32*, 285–309. [CrossRef]
36. Almstedt, Å.; Brouder, P.; Karlsson, S.; Lundmark, L. Beyond Post-Productivism: From Rural Policy Discourse to Rural Diversity. *Eur. Countrys.* **2014**, *6*, 297–306. [CrossRef]
37. Murdoch, J.; Pratt, A.C. Rural studies: Modernism, postmodernism and the "post-rural". In *The Rural*; Munton, R., Ed.; Routledge: London, UK, 2016. [CrossRef]
38. Oliva, J.; Camarero, L.A. Shifting rurality: The Spanish countryside after de-peasantisation and de-agrarisation. In *Europe's Green Ring*; Granberg, L., Kovách, I., Tovey, H., Eds.; Routledge: London, UK, 2001. [CrossRef]
39. Tilzey, M.; Potter, C. Governance in post-Fordist agricultural transitions. In *Sustainable Rural Systems: Sustainable Agriculture and Rural Communities*; Robinson, G., Ed.; Routledge: London, UK, 2008. [CrossRef]
40. Vidickienė, D.; Vilkė, R.; Gedminaitė-Raudonė, Ž. Transformative Tourism as an Innovative Tool for Rural Development. *Eur. Ctries.* **2020**, *12*, 277–291. [CrossRef]
41. Gómez y Patiño, M.; Medina, F.X.; Puyuelo Arilla, J.M. New trends in tourism? From globalisation to postmodernism. *Int. J. Sci. Manag. Tour.* **2016**, *2*, 417–433.
42. Băndoi, A.; Jianu, E.; Enescu, M.; Axinte, G.; Tudor, S.; Firoiu, D. The Relationship between Development of Tourism, Quality of Life and Sustainable Performance in EU Countries. *Sustainability* **2020**, *12*, 1628. [CrossRef]
43. Fialová, D.; Chromý, P.; Kučer, Z.; Spilková, J.; Štych, P.; Vágner, J. The forming of regional identity and identity of regions in Czechia–Introduction to the research on the impact of second housing and tourism. *Acta Univ. Carol. Geogr.* **2010**, *45*, 49–60. [CrossRef]
44. Šťastná, M.; Vaishar, A.; Pákozdiová, M. Role of tourism in the development of peripheral countryside. Case studies of Eastern Moravia and Romanian Banat. *Forum Geogr.* **2015**, *14*, 83–93. [CrossRef]
45. Konečný, O. Geographical perspectives on agritourism in the Czech Republic. *Morav. Geogr. Rep.* **2014**, *22*, 15–23. [CrossRef]
46. Kogler, D.F.; Whittle, A. The geography of knowledge creation: Technological relatedness and regional smart specialization strategies. In *Handbook of the Geographies of Regions and Territories*; Paasi, A., Harrison, J., Jones, M., Eds.; Edward Elgar: Cheltenham, UK, 2018; pp. 153–168.
47. Mardani, A. Evaluation of quality of life in rural areas, case: Sadeghieh rural district of Najafabad. *Geogr. Hum. Relatsh.* **2021**, *3*, 408–418. [CrossRef]
48. Kebza, M. The development of peripheral areas: The case of West Pomerania voivodeship, Poland. *Morav. Geogr. Rep.* **2018**, *26*, 69–81. [CrossRef]
49. Hoogerbrugge, M.; Burger, M. Selective migration and urban–rural differences in subjective well-being: Evidence from the United Kingdom. *Urban Stud.* **2021**. [CrossRef]
50. Kubeš, J. Chatové oblasti České republiky. *Geogr. Časopis* **2011**, *63*, 53–68.
51. Curta, F. The history and archaeology of Great Moravia: An introduction. *Early Mediev. Eur.* **2009**, *17*, 238–247. [CrossRef]
52. Kolejka, J.; Nováková, E.; Kirchner, K. Residual ancient landscape segments in Carpathians of Moravian-Slovak borderland. In *Public Recreation and Landscape Protection with Sense Hand in Hand*; Fialová, J., Ed.; Mendel University: Brno, Czech Republic, 2020; pp. 408–415.
53. Urbanová, M.; Dundelová, J. Work culture of the Bata company. *Acta Univ. Agric. Silvic. Mendel. Brun.* **2012**, *60*, 487–494. [CrossRef]
54. Skřivan, A., Jr. On the Nature and Role of Arms Production in Interwar Czechoslovakia. *J. Slav. Mil. Stud.* **2010**, *23*, 630–640. [CrossRef]
55. Řehák, S. The Moravian-Slovak borderlands: Some new features following the division of Czechoslovakia. *Morav. Geogr. Rep.* **1998**, *6*, 14–17.
56. Halás, M. Theoretical preconditions versus real existence of cross-border relations in the Czech-Slovak borderland. *EUROPA XXI* **2006**, *15*, 63–75.
57. Suomalainen, J. Reforming Borderlands in the European Union. Understanding the Czech-Slovak Borderland Dynamics. Master's Thesis, University of Tampere, Tampere, Finland, 2018.
58. Oremusová, D.; Kramáreková, H.; Nemčíková, M. Evaluácia projektových aktivít mikroregiónu Termál v troch programových obdobiach. In *24. Mezinárodní Kolokvium o Regionálních Vědách*; Klímová, V., Žítek, V., Eds.; Masarykova Univerzita: Brno, Czech Republic, 2021; pp. 108–116. [CrossRef]
59. Pavlakovich-Kochi, V.; Morehouse, B.J. *Challenged Borderlands. Transcending Political and Cultural Boundaries*; Routledge: London, UK, 2004. [CrossRef]

MDPI
St. Alban-Anlage 66
4052 Basel
Switzerland
Tel. +41 61 683 77 34
Fax +41 61 302 89 18
www.mdpi.com

Sustainability Editorial Office
E-mail: sustainability@mdpi.com
www.mdpi.com/journal/sustainability

www.ingramcontent.com/pod-product-compliance
Lightning Source LLC
LaVergne TN
LVHW070902170726
843515LV00005B/693

* 9 7 8 3 0 3 6 5 4 4 7 9 3 *